TERTIARY LEVEL BIOLOGY

An Introduction to Coastal Ecology

PATRICK J.S. BOADEN, B.Sc., Ph.D.

Director
Marine Biology Station
Queen's University, Belfast

RAYMOND SEED, B.Sc., Ph.D.

Senior Lecturer
School of Animal Biology
University College of North Wales

WITHDRAWN

Blackie

Glasgow and London
Distributed in the USA by
Chapman and Hall
New York

Blackie & Son Limited,
Bishopbriggs, Glasgow G64 2NZ

Furnival House, 14–18 High Holborn, London WC1V 6BX

Distributed in the USA by
Chapman and Hall
in association with Methuen, Inc.
29 West 35th Street, New York, N.Y. 10001
© 1985 Blackie & Son Ltd
First published 1985

British Library Cataloguing in Publication Data

Boaden, Patrick J.S.
An introduction to coastal ecology.—(Tertiary level biology)
1. Coastal ecology
I. Title II. Seed, Raymond III. Series
574.5′2636   QH541.5.C65

ISBN 0–216–91795–6
ISBN 0–216–91796–4 Pbk

Library of Congress Cataloging in Publication Data

Boaden, Patrick J.S.
An introduction to coastal ecology.

(Tertiary level biology)
Bibliography: p.
Includes index.
1. Coastal ecology. I. Seed, Raymond. II. Title.
QH541.5.C65B63 1985   574.5′2638   85-10937
ISBN 0–412–01021–6
ISBN 0–412–01031–3 (pbk.)

Photosetting by Thomson Press (I) Limited, New Delhi
Printed in Great Britain by Bell & Bain (Glasgow) Ltd.

Preface

Studies of marine ecology have traditionally been approached through lectures and field courses devoted mainly to intertidal and inshore habitats, and it is surprising in these days of increased awareness of man's environmental impact that so little attention has been given to integrated approaches involving the whole coastal zone and including the terrestrial part, which is man's major habitat. The coastal zone has been the subject of extensive investigation, not only because of its biological diversity and accessibility, but also because of its economic and aesthetic importance to man.

This book is written with the intention of providing a concise but readable account of coastal ecology for advanced undergraduates and immediate postgraduates. We have adopted a habitat–organismal approach because we believe that a knowledge of biota and major features of their environment is the best key to an understanding of both larger-scale processes, such as energy flow and nutrient cycling, and smaller-scale but equally fundamental processes, such as behavioural and physiological ecology. Examples have been selected from polar, temperate and tropical regions of the world. The breadth of the subject has dictated selectivity from sources too numerous to acknowledge individually, but we have included an up-to-date reference list for the main subjects of each chapter.

We would like to thank the many people who have helped in the production of this book. As authors, we jointly shoulder responsibility for the contents, style and accuracy, but we wish to acknowledge all those who have given physical, moral and intellectual support, including staff in our own and other Departments of the University College of North Wales and the Queen's University of Belfast, and especially Drs R.W. Arnold, A.J. Butler, C.J. Feare and G. Savidge.

For Cherry and Wendy

Contents

Distribution and limiting factors. Reef morphology and zonation. Nutrition, calcification and growth. Species interactions—mutualism, competition, predation, grazing. Physical disturbance. Diversity gradients and biogeography. Productivity.

Saltmarshes: Establishment. Succession, zonation and marsh structure. Other flora and fauna. Trophic structure and energy flow. **Mangroves**: Geographical distribution. Zonation. Succession. Adaptation of mangroves. Mangrove fauna. Primary productivity; trophic interrelationships.

Sand dunes—the strandline, dune ridge formation, dune slacks, dune fauna. West coast deserts. Shingle beaches. Grassland and heath. Coastal woodlands. Cliffs. Swamps and mires. The ice edge.

Sponges. Corals. Worms. Crustacea—shrimps, crabs, lobsters and spiny lobsters. Mollusca—bivalves, gastropods, squid and octopus. Commercial fish—demersal fish, pelagic grazers, pelagic predators, fisheries research, fish culture. Other coastal fish. Higher marine vertebrates—snakes and lizards, crocodiles and turtles, marine mammals. Seaweed resources. Environmental impact of fisheries.

Classification of coastal birds. Distribution. Adaptations to the marine environment. Food acquisition. Breeding ecology. Mortality, longevity and regulation of numbers. Migration and foraging. The importance of birds in coastal ecosystems.

Physical resources—space, power generation. Mineral resources. Pollution—pesticides and related compounds, heavy metals, radioactivity, oil, organic matter, nutrients. Public health. Pests and introductions. Fouling. Access, recreation and tourism. Planning and legislation. Coastal conservation—detection and prediction of change, coastal nature reserves. System modelling. In conclusion. Envoi.

CHAPTER ONE

THE COASTAL ENVIRONMENT

It is the purpose of this book to present a general account of the ecology not only of the coast itself—that is the land next to the sea including the intertidal zone—but also of its adjacent coastal waters. In particular it is hoped to introduce the reader to coastal phenomena and their underlying causes through description and explanation of observable physical and biological features of the environment.

Coastal areas can be regarded as the interface between three habitable media, namely earth, air and sea. However, this interface is composed of a series of boundaries or gradients whose dimensions range from a few nanometres or less, for example a water film around a sand grain, to a few kilometres or more, for example the windborne carriage of sea-spray inland. These boundaries are not constant but change in space and time, sometimes slowly, sometimes rapidly. Some changes such as tidal rise and fall are predictable and regular, others stochastic, that is more or less unpredictable or at random.

All coastal areas contain at least two obvious habitats, namely the *maritime zone*, which is home for many terrestrial animals and plants, and the *sea* itself. Although there are several areas in the world where water depths greater than 200 m are found close inshore, this book will restrict its marine sections to consideration of the ecology of the shallow-water zone overlying the continental shelf. A third obvious habitat, the *intertidal zone*, sometimes called the *littoral* (see Chapter 3), occurs along much of the world's coastlines, but even in areas without noticeable tides there is often a specialized fauna and flora at the sea–land interface.

Coastal topography

Neither the sea nor the coastline are static. Occasionally there are dramatic changes due to vulcanism or severe meteorological conditions. The volcanic destruction of the island of Santorini, or more recently, Krakatoa

1

(1883), the 1963 creation of Surtsey off Iceland or the intermittent appearance of Falcon Island in the Pacific are extreme examples. Landslides, such as at Dowlands, England, on Boxing Day 1839 (see Steers, 1964), and gale erosion may also cause sudden substantial change. However, most coastal realignment is a gradual process. The outline of a natural coastline is determined by its geological composition and structure, by terrestrial weathering and drainage, by inshore topography and hydrography and by the relative level of water to land. There are three basic types of coastline.

Embayed coasts

These are generally rocky and characterized by extensive inlets, bays or fjords. Examples are to be found in north-west Europe, Chile, Alaska and the north-eastern shores of North America and New Zealand.

Plains coasts

These are low-lying sand and mud coasts, generally with offshore sandbars and islands. A classic example is the Cape Kennedy or Hatteras coastline of the eastern United States.

A broadly similar classification is that used by Steers (1964) and other geographers which distinguishes between submerged and emerged shorelines. Submerged shores have resulted from the land sinking relative to sea-level and thus producing a typically embayed coastline. Lifting of the sea-floor or a fall in sea-level tends to produce a plains coast. Vertical movements of the entire land mass or of sea-level are referred to as *isostatic* and *eustatic* changes respectively; where the sea-level rises relative to land the movement is termed positive; when it becomes lower in comparison, negative.

New coasts

These are areas where material has been added by sedimentation and accretion, by vulcanism or by organic processes such as coral growth to produce a new shoreline without the necessity of isostatic or eustatic movement.

Various attempts have been made to classify coastlines on a worldwide basis (Strahler, 1963). Some common types are shown in Figure 1.1.

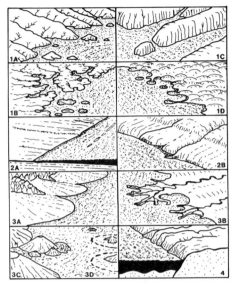

Figure 1.1 Major varieties of shorelines. 1*A*, Ria shoreline; 1*B*, shoreline of submergence with low relief; 1*C*, fiord shoreline; 1*D*, drumlin shoreline. 2*A*, coastal-plain shoreline; 2*B*, shoreline of emergence, steeply sloping. 3*A*, alluvial-fan shoreline; 3*B*, delta shoreline; 3*C*, volcano shoreline; 3*D*, coral-reef shoreline. 4, fault shoreline. (From Strahler, 1960.)

Marine biologists tend to make their initial binary distinction on the basis of shores being composed of either hard (rock, boulder and cobble) or soft (gravel, sand and silt) substrata. Their secondary distinction is usually whether the area is exposed or sheltered regarding wave action.

Cliffs

Cliffs are a common feature where rock meets sea. Often the cliffs are underlain by a wave-cut platform. Extensive platforms occur where the rock is soft (e.g. limestone) and where the land–sea levels have been relatively static especially after earlier marine transgressions or coastal submergence. Beaches beneath cliffs absorb a large part of the incoming wave energy and thus decrease platform formation. Cliff morphology may be influenced by the bedding or strata. Where these are horizontal, cliffs tend to be vertical. If strata dip seaward, the upper cliff slope may be greater than, or overhang, the lower slope. Strata sloping down landwards tend to produce gentler slopes. Folded or perpendicular strata often give rise to complex local indentations. The erosive forces due to wave action (including compression and expansion of trapped air), tidal scour, dissolution and frost act on joints or cracks and on weaker materials, sometimes causing fairly large blocks of material to fall. Ice is an important erosive agent in high latitudes. Cliffs formed from glacial deposits or other unconsolidated material are prone to extensive erosion by waves, currents and heavy rain; wind can pluck material from such cliff faces and slipping may be caused by percolation of groundwater especially where heavier material overlies clay.

Sediments

Material eroded from the coastline or brought to the sea from other sources is likely to be transported elsewhere, either immediately or after further breakdown. The movement of intertidal material can be brought about by wind or fresh water, but normally transport of material here and in shallow water is the result of the interaction of particle size, and to some extent shape, with the marine hydrodynamic regime (see Allen, 1970).

Theoretically, small (< 0.2 mm diameter) spherical particles should sink through water at a *settlement velocity* (Figure 1.2) determinable through

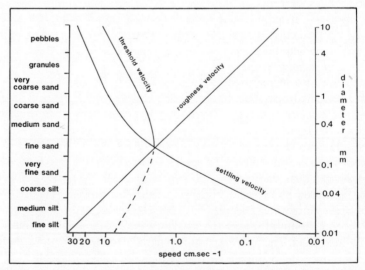

Figure 1.2 Settlement and stability of bottom particles. For further details see text. (After Inman, 1949.)

Stoke's Law which states that the terminal velocity of settlement will be inversely proportional to fluid viscosity and directly proportional to the square of the particle diameter. Particles larger than 2 mm settle at a rate proportional to their diameter's square root. Between these sizes rates are intermediate. However, under natural conditions there will generally be some turbulence in the water column and the vertical components may accelerate or slow settlement: horizontal currents will determine the total distance travelled before the particle reaches the bottom. Obviously deposition will be greatest where there is a steady supply of materials into relatively still water. Postma (in Olson and Burgess, 1967) lists three types of inshore area where accumulation of suspended sediment is likely to be high. These are: (1) beaches—due to breaking waves onshore water movement is faster though of shorter duration than offshore movement, hence particles tend to have net shoreward movement; (2) estuaries—due to sediment accumulation in the salt wedge (Chapter 5); (3) tidal flats—currents further out to sea tend to be stronger, especially between sandbanks and islands where the water is also deeper, and material brought closer inshore is subject to weaker currents and has less distance to settle.

Once material has been deposited it may be covered by further particles and eventually enter into the sedimentary rock-forming process known as diagenesis. However, due to bottom current increase, albeit temporary, particles may be rolled or carried along the sediment surface (bed load transport) or carried back into the overlying water (resuspension). There are two critical current speeds along the bottom, again related to particle size, above which these processes occur (Figure 1.2). These are in the first instance *threshold velocity*; this increases for large particles because of inertia but also increases for small particles because fine sand and silt surfaces are hydrodynamically smooth. The critical resuspension factor is *roughness velocity*, that is the current speed over the particles at which flow ceases to be laminar. This is inversely related to particle size.

Natural features may be created or changed by deposition and accretion. Among these are sand dunes, shingle banks and sand beaches (Figure 1.3). Material can be transported along beaches due to the oblique approach of waves followed by more vertical backwash. This process is called beach drifting. Inshore water may also have currents brought about by a head of water being produced against the coast by wave mass transport, often aided by onshore wind. The water then moves parallel to the shore towards areas with less pressure. Sediment carried in these longshore currents produces longshore drift.

Where there are offshore sandbanks, bars or islands strong tidal currents

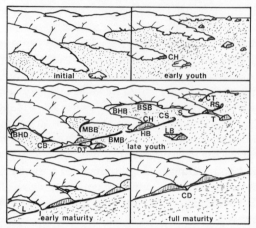

Figure 1.3 Evolution of an embayed coast. BHB = bayhead beach; BHD = bayhead delta; BMB = bay-mouth bar; BSB = bayside beach; CB = cuspate bar; CD = cuspate delta; CH = cliffed headland; CS = cuspate spit; CT = complex tombolo; DT = double tombolo; HB = headland beach; I = inlet; L = lagoon; LB = looped bar; RS = recurved spit; S = spit; T = tombolo. (From Strahler, 1960.)

moving directly inshore or offshore often exist. Among the most notorious are sand beach rip currents which reach 4–5 km h^{-1} and have been responsible for the death of many bathers. Currents of similar or greater strength can occur in rock and coral gullies. Subtidal sand (and shingle) is sometimes deposited in interdigitating banks. In areas of relatively strong bottom currents, such as the western English Channel, sand waves (slowly moving submarine dunes) and ribbons may be formed even at depths of 150 m. Muddy sediments accumulate in quieter water. Sediment-laden water acts as a denser fluid than 'clean' sea water and can give rise to severe turbidity currents as it pours into shelf canyons or over the shelf edge.

Water movement

It should be clear from the foregoing that water movements play a key role in coastal topography. They are also highly important in marine ecology. Coastal environments are affected by five main types of seawater movement, namely waves, tides, seiches, currents and aerial transport of spray. Movement of fresh water by evaporation, precipitation, run-off and drainage also has coastal effects but these will be evident from later chapters.

Waves

Although the waves usually seen at coasts are wind-generated, four other types—internal waves, storm surges, tsunamis, and tides—also occur (see Fairbridge, 1966, or Turekian, 1976). Basic wave form is shown in Figure 1.4. Within waves, water particles are in orbital motion, although there is some slight forward motion (mass transport) proportional to the square of the face steepness. At the surface, the orbit diameter equals wave height but the orbit lessens with depth (halving with each ninth of wavelength depth). However, immediately next to the shore orbital motion will be restricted and there will be multidirectional movement to depths equivalent to 2.5 wave amplitudes. From here to depths corresponding to 0.5 wavelengths there is an oscillating surge parallel to the sea-bed slope. Below this there is a unidirectional residual current (Figure 1.4). In soft sediments water may be driven through surface sand by wave action.

The dimensions of wind-generated waves are determined by the wind's speed, duration and *fetch*. Fetch is the uninterrupted distance across water over which the wind has acted. In areas of actual wave generation, wave patterns are very complex and are termed a *sea*. In a sea, waves have a fairly wide spectrum of amplitudes, periods and directions. Normally one wave

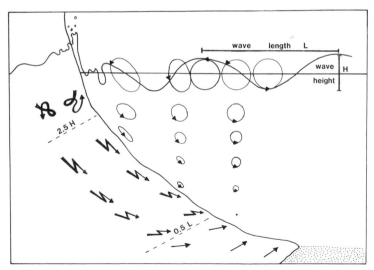

Figure 1.4   Wave-induced subtidal currents on a rocky shore. Unpredictable water movement is replaced by oscillating currents and then by tidal or other laminar flow with depth. (After Riedl, 1971.)

period and height becomes dominant. Thus in the North Atlantic, force 4 winds generate waves with a dominant 6-second period and 0.6 m amplitude, whereas force 8 winds give an 11-second interval and 6.0 m height. After generation, wave motion spreads outwards downwind and to the sides of the generating fetch. The outward-travelling waves, often called *swell*, lose height since energy per unit horizontal length of wave force is spread out as the wave front widens in the open ocean. However, wave length remains constant. The series of waves travelling away from the sea forms a wave train which advances at half the speed of the apparent individual wave movement. The rate of height-decrease lessens as the wave train advances; just over half the height is lost after 2500 km travel. Because of their immense initial energy, hurricane-generated waves can cross entire oceans. For example, those generated off the eastern United States can reach western Europe after 4–5 days.

As waves move toward coasts, frictional forces beneath the wave come into effect once the sea depth becomes less than half the wave length. Wave period remains the same but height and steepness increase whilst wave length is reduced. The bottom friction causes wave refraction since the shallower parts of the wave are slowed more rapidly than deeper parts. Thus the wave fronts slew, tending to become parallel to the depth contours. This means that wave energy is concentrated towards promontories and over subtidal ridges but correspondingly lessened elsewhere (Figure 1.5). If wave height reaches three-quarters of the water depth,

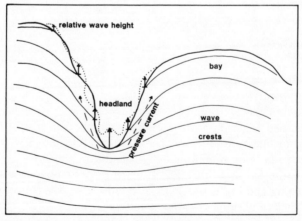

Figure 1.5 Wave alignment on coasts. Wave energy is concentrated on headlands but dissipated in bays.

either by the wave steepening or by the bottom shallowing, the wave becomes unstable and breaks, as much of its energy is translated into forward movement. On sloping shores the broken wave runs up the beach as swash and then down as backwash; these are sometimes called uprush and backrush. On vertical shores air trapped as the wave topples may explode through the wavetop, thereby producing a sheet of spray, some of which may be carried far inland by wind. Where waves are fairly constantly experienced they effectively raise the intertidal level at which marine organisms can survive. This effect is proportionally greater in areas with a small tide range.

It appears that normal winds are also connected with *surf-beat*. These are waves with a period about ten times longer than ordinary waves and which possibly account for the widespread folk belief that every seventh wave is larger. Surface ripples occur on the sea surface even in relatively calm conditions. Although they are normally wind-generated due to turbulent air flow and consequent pressure differences at the interface, their size is controlled by water density and surface tension. Ripples are important in aiding wave generation since they give a rougher and greater surface area for the wind to act on. In sheltered conditions, smooth rather glossy patches, known as slicks, are often interspersed with rippled areas. Stronger wind elongates or breaks up the slicks into narrow bands called windrows. Slicks are due to concentration of natural or anthropogenic organic material sufficient to form a surface monomolecular layer. Such material is probably involved in stabilizing surface bubbles to produce foam.

Internal waves

Waves can be generated within the overall water column where vertical density gradients or discontinuities exist. They may result from various causes, such as shear force between water layers or from flow over an uneven sea-bed. In deep water the amplitude can be considerable, but in the main thermocline (p. 15) the waves are usually 6 – 10 m high. However, the surface and sea-floor amplitudes are negligible or zero and the waves are essentially internal structures. The water displacements caused are important in mixing and in the transport of solutes. Over the continental shelf, bands of slicks and ripples parallel to the coast can be taken as evidence of internal waves, the slicks indicating underlying wave back-faces and the maximum ripple areas indicating the internal wave crests.

Storm surges

Changes in barometric pressure cause fluctuations in mean sea-level, a pressure increase of 1 mm Hg causing a downward movement of 13 mm or a 1 millibar increase a 9.75 mm drop. A change between strong cyclonic and anticyclonic pressure will produce a sea-level change of about 50 cm. However, when a storm passes across shallow water, not only will wave height increase and sea-level rise, but if the storm speed-of-passage matches wave speed a resonance is produced. This may raise the water height by over 6 m, resulting in a greatly steepened large wave known as a storm surge. The effects of storm surges on the coast may be just as severe as those of tsunamis. The Ganges delta has repeatedly suffered from catastrophic surges generated by Bay of Bengal cyclones. Storms crossing the North Sea from north-west to south-east can generate surge waves 2–3 m above predicted tidal level; they travel anticlockwise around the North Sea but fortunately their arrival does not coincide with high tide on adjacent coasts. In early 1953 over 2000 people were drowned in England and the Netherlands following a severe surge.

Tsunamis

Sudden displacement of large volumes of solid material caused for example by landslides, earthquakes or volcanic eruption underwater sets very fast-moving waves in motion. The effect could be likened to 'dropping' a giant pebble into water from underneath. The waves spread outward like ripples. In open ocean, their amplitude may be less than 1 m but the wave length as long as 250 km and the wave period 10–30 minutes. The apparent forward wave motion given a 200 km wave length and a 20-minute period would be 600 km h^{-1}. Frictional forces are so great when entering shallow water that wave height may increase to 10 m or more before the tsunami (and its surface wind waves) breaks into cataclysmic coastal surf. It is wrong to call these waves, such as followed the Lisbon earthquake of 1755, or storm surges 'tidal waves' since they have very different causes and consequences than true tides.

Tides

Tides are waves of approximately 12.5 h or 1-day period caused in our planet's seas as an outcome of solar system physics—chiefly the gravitational effects of the sun and moon combined with effects of the earth's rotation. Additional longer-term cycles are superimposed through the effects of the earth's varying declination and the orbital paths of the earth and moon. Tides are also affected by topographical features, for example estuaries and bays can produce amplification and there may be interaction with seiches (see later). Atmospheric pressure and wind conditions also modify tidal levels.

Consider the earth in orbit around the sun but initially neglect the earth's rotation around its axis (Figure 1.6). The average gravitational pull, tending to draw the earth and sun together, must be matched by the average centripetal acceleration tending to move the earth outwards—otherwise the earth's orbit would change. However, the actual gravitational force at a point P will vary with place and time from that at the earth's centre C. The tide force F is the difference between the actual and average forces and approximates to

$$F_S = GM\frac{2a}{r^3}$$

where a is the difference in distance of C and P from the sun, r is the distance of C from the sun, G is the universal gravitational constant and M the sun's mass. A similar calculation can be made in respect of the moon (producing F_L) although it is revolving around the earth. Because of its proximity the moon's gravitational effect is twice that of the sun.

The resultant total tide force F_T will vary since F_S and F_L may act in the same plane (when moon, earth and sun are aligned) or diverge. Alignment occurs one and a half days after new and full moon, producing the large

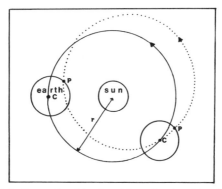

Figure 1.6 Eccentric orbit of earth's surface localities. See text.

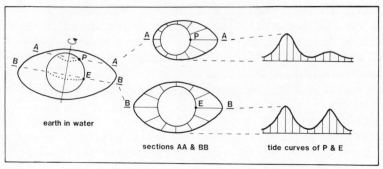

Figure 1.7 Tidal analogy. The rotation of the earth beneath a varying thickness of sea produces near-equal or unequal semi-diurnal tides of E and P respectively.

tidal range known as a *spring* tide. Seven days later the forces are 90° apart producing the small range of *neap* tides. At the equator F_T is directed nearly vertically upward toward the moon but at the poles downward toward the earth's centre; since the earth's own gravitational pull on the sea is 10^7 greater than F_T there can be virtually no lateral tidal displacement at these latitudes. However at 45° N and S lateral displacement is at a maximum since F_T acts tangentially to the earth.

An exaggerated analogy to tides on a completely water-covered earth is a static rugby ball containing a smaller spinning football. The latter represents the solid earth and the space between the two balls represents a world ocean (Figure 1.7). In spinning round once (a day), a point near the equator of the football will pass nearly beneath a point of the rugby ball, then beneath its smallest central diameter, then beneath the second point and then again under the smallest diameter. These four instances represent the two high waters and low waters which occur semidiurnally as the result of the tide wave which has a period of 12 h 25 min and a length of half the appropriate latitudinal perimeter of the earth. However if a point at 45° N or S is imagined similarly it is clear that the two high tides will be of very unequal dimensions; increasing this inequality by tidal basin and other effects leads to the concept (and reality) of diurnal tides with a lunar day period and a latitude-perimeter wave length.

Land mass and sea-bed shape cause the sea to be divided into a series of basins each with its own tidal characteristics. Geostrophic forces cause the tide in each basin to rotate about a node, termed the *amphidromic point*, at which the tide range is zero (Figure 1.8).

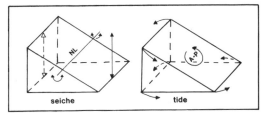

Figure 1.8 Model of a seiche oscillating about a nodal line (*NL*) and a tide rotating around an amphidromic point (*AP*).

Seiches

Any basin of water, whether a harbour, a semi-enclosed sea or a plastic bowl, has a natural resonance. On momentary disturbance the water will oscillate with decreasing amplitude until becoming still. Such natural oscillations are termed seiches (Figure 1.8). Tides in adjoining basins or tides interacting with seiches may complement or cancel each others' effect, thus there may be considerable differences in tide range along the neighbouring coasts. Coastal configuration may also increase or decrease tidal amplitude; the best known examples are in deep embayments and estuaries, such as the Bay of Fundy, where energy concentration and tidal resonance produces spring tides of up to 17 m.

In the intertidal zone the length of time any part is covered (*submersed*) or uncovered (*emersed*) by the tide varies with its shore height (Figure 1.9). If the tide curve is symmetrical mid tide level (MTL) will be covered and uncovered for equal periods of time. Areas below the level of the least high tide (extreme-low high water neap E(L)HWN) will be covered for at least a short period in every tide cycle, but above this level areas may be uncovered for one or more days. Above high water spring tide level (HWS) the shore may be emersed for a fortnight or more. The reverse situation applies to submersion or to low shore levels.

Currents

Currents are the fourth main type of water movement. Coastal waves and tides can both produce currents; often in the latter case the direction reverses with ebb and flow but there is usually an overall (residual) movement in one direction. Two other forces are of prime importance in producing coastal currents. These are *horizontal drag* from wind and

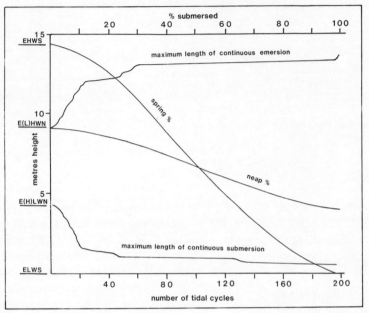

Figure 1.9   The relation between height on the shore, the total percentage submersion and the maximum period of continuous emersion or submersion. Data from Avonmouth, England. (After Dring, 1982.)

vertical momentum due to density differences. The latter can arise from temperature or salinity changes. The major ocean currents also affect adjacent land, mainly via the atmosphere, through heat transfer, evaporation and precipitation. As a gross generalization major winds tend to blow parallel to the coasts. Wind-drag generated currents tend to move at an angle (to the right in the Northern Hemisphere, left in the Southern) away from the wind destination. Surface water moved away from the shore is replaced by colder and generally nutrient-rich water from below (Figure 1.10), hence the fauna and flora of such areas is often abundant. Such *upwelling* currents occur off many subtropical west coasts. The cooling effect may produce coastal fogs as found off the western United States. Coastal zone *fronts* (that is boundary areas between different coastal water masses with different physical and chemical characteristics) are also highly important in upwelling, mixing and overall productivity. Upwelling results in divergences of surface water. Downwelling or *convergence* may also occur at fronts or along coasts. Turbulence and eddy diffusion are also important mixing processes.

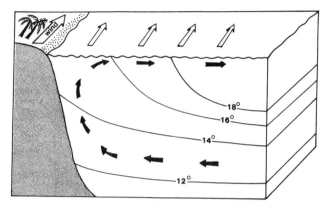

Figure 1.10   Wind-induced coastal upwelling. Solid arrows indicate subsequent currents, open arrows the surface water movement. (After Fairbridge, 1966.)

Temperature, salinity and thermoclines

Where strong mixing forces do not exist, fairly discrete bodies of water with their own physical, chemical and biological characteristics are often recognizable. Both temperature increase and salinity decrease lower the density of sea water, thus a surface water layer can float on a colder or more saline layer. Such surface layers can become depleted of nutrients by biological activity and the deeper layer can become oxygen-depleted. Where such horizontal boundaries exist due to temperature difference and column stability they are called *thermoclines*. These are often seasonal or transient in coastal waters, and tend to be close to the surface in the tropics but deeper in high latitudes. The surface temperature of sea water usually lies between -2 and $+30°C$. Salinity, usually written as $S‰$ (parts of salt per thousand of water), in the open oceans is usually between $33-37‰$ but $33-34‰$ may be taken as a more usual coastal value. In parts of the inner Baltic, salinity is less than $5‰$, whereas in enclosed evaporative areas such as the Red Sea the salinity may exceed $40‰$. The major salt constituents in sea water have a fairly constant concentration relative to each other. However, some of the minor constituents, such as nitrate, phosphate and silicate vary markedly with considerable biological consequences (Chapter 2). Among the dissolved gases oxygen is of particular biological importance and its concentration is linked to primary production and respiration. Surface water may become supersaturated ($< 130\%$) or entirely depleted of oxygen—normal coastal values are 4–8 ml

$O_2 1^{-1}$. Carbon dioxide concentration is more constant since it is effectively 'buffered' by the carbonate alkalinity system of sea water (see Hill, 1962; Goldberg, 1974).

Organic matter

In addition to its inorganic major, minor and trace constituents sea water contains both particulate and dissolved organic matter. A detailed account will be found in Riley (1970). Particulate organic matter (POM) as measured in the laboratory normally includes some bacteria and phytoplankton, but the larger part is composed of 'detritus'. POM with its attached micro-organisms is an important food for filter feeders, although the detrital particles may contain a considerable proportion (70%) of inorganic matter. POM is also important in surface slick formation. Dissolved organic matter (DOM) is usually distinguished from POM by its ability to pass through very fine filter pores and therefore is not necessarily entirely dissolved. It contains many of the same substances as POM including amino acids and carbohydrates. It is known that many marine organisms can make use of these compounds by active transport through their epidermal surfaces (Stewart, 1979).

Maritime climates

The climate next to the sea differs from that further inland. The coast often receives stronger predominant winds due to lack of shelter and there may also be the well-known land and sea breezes.

Maritime areas are often wetter than inland due to their proximity to the world's main source of evaporation (the oceans) and to arographic or to frontal climatic effects, that is the wind – forced upward ascent of air by increased land altitude or the rising of warm air on meeting a cooler air mass respectively. Maritime lands are often well watered by the lower reaches of rivers and the surface proximity of groundwater. In peninsulas and islands of suitable structure, groundwater exists as a convex lens, the thickest part lying about forty times deeper than the maximum water table height lies above sea-level. However, coastal deserts also exist for example in Chile, Namibia and north-western Australia. These west coast deserts are cooled by proximity to major upwelling currents. Conversely, maritime zones may be warmed by heat from warm currents. An example is the influence of air warmed by the Gulf Stream on the western parts of northern Europe. Similarly extension of warm climate vegetation to higher latitudes occurs on the eastern coasts of northern Australia and Sri Lanka, of southern Brazil and of Panama and Nicaragua (Strahler, 1963). Maritime lands often have an equable climate due to the heat storage effect of the adjacent sea.

It can be argued that marine and terrestrial coastal habitats are generally of higher primary productivity (see p. 33) than those inland or offshore due to the availability of water and nutrients. The transfer of salt to land or of fresh water to sea can cause environmental stress. Many coastal environments experience relatively rapid change which may be cyclic or irreversible; however, these changes can lead to resource renewal, for example, by producing fresh substratum or by recycling nutrients. Such stress and flux in the environment will favour some species because of their evolutionary specialization or because their competitors are less effective or absent in such conditions.

CHAPTER TWO

THE NERITIC PROVINCE

The world's seas can be broadly divided into *benthic* (sea-bed) and *pelagic* (water-column) realms both of which can be further subdivided according to depth (Figure 2.1). The pelagic realm consists of *oceanic* and *neritic* provinces, the latter comprising the shallow (down to 100–200 m) coastal waters overlying the continental shelf. Although accounting for only a comparatively small proportion (*c.* 5%) of the world's seas, neritic environments are often highly productive and many of the important commercial fisheries are located in these inshore waters. The *euphotic* (= epipelagic) zone is that part of the pelagic realm which is well illuminated. Its lower boundary varies according to the clarity of the water but is usually around 100–200 m. It grades into the *aphotic* zone which is permanently dark below about 1000 m.

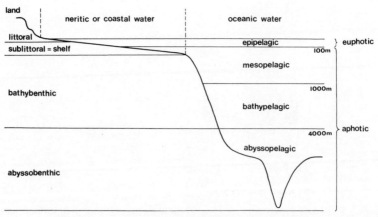

Figure 2.1 Major subdivisions of the marine environment.

Plankton

Irrespective of phylogenetic position two major categories of pelagic organisms can be recognized: *plankton* and *nekton*. A third category, the *pleuston*, comprises those organisms that straddle the air–water interface (e.g. the siphonophore *Physalia* and the gastropod *Ianthina*—the former using a gas-filled float, the latter a raft of air bubbles).

Planktonic organisms (Figure 2.2) have limited powers of locomotion, at least horizontally, and consequently they are more or less passively drifted by the water currents. They include plants (*phytoplankton*), animals (*zooplankton*) and bacteria (*bacterioplankton*). The distinction between the phytoplankton and the zooplankton, however, is not always clearly defined, since some species live as autotrophs in the light and heterotrophs

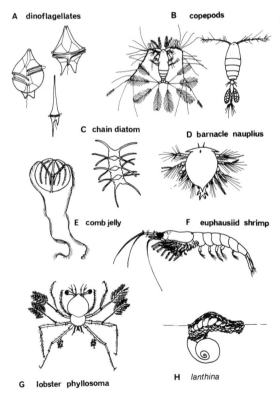

Figure 2.2 Some representative members of the plankton (redrawn from various sources). Not to scale.

Table 2.1 Classification of plankton by size

Type	Size range	Examples*
Femtoplankton	0.02– 0.2 μm	Viruses
Picoplankton	0.2– 2.0 μm	Bacteria
Nanoplankton	2.0– 20 μm	Small autotrophic flagellates
Microplankton	20– 200 μm	Protozoans, diatoms, dinoflagellates
Mesoplankton	0.2– 20 mm	Copepods
Macroplankton	2.0– 20 cm	Krill, arrow-worms
Megaplankton	0.2– 2.0 m	Large jellyfish

*There is often overlap between neighbouring categories.

in the dark. Classification of plankton is often based on size (Table 2.1) or life-history characteristics. Members of the *holoplankton* spend their entire life in the plankton, whilst members of the *meroplankton* are temporarily planktonic. The *neuston* consists of very small planktonic organisms (e.g. bacteria, protozoans) found within an ultra-thin layer at the sea surface. This subsurface film is particularly rich in salts and dissolved organic material (DOM). Many small benthic organisms may be temporarily whirled up off the bottom by currents and turbulence; these are collectively termed the *tychoplankton*. The distinction between even planktonic and benthic organisms is thus perhaps much less clearcut than might be presupposed.

Larger pelagic organisms (e.g. fish, cephalopods, mammals) that are capable of counteracting water currents by swimming constitute the *nekton* (see Chapter 9).

Phytoplankton

The phytoplankton is especially important in marine food chains, as these organisms are the primary producers (autotrophs), building up complex organic molecules directly from carbon dioxide and inorganic nutrients (e.g. phosphorus, nitrogen) present in the sea. Diatoms are non-motile unicellular algae which occur either singly (e.g. *Coscinodiscus*) or as chains (e.g. *Chaetoceros*). All have two silicified valves (the frustule) which fit together like a pill box. Diatoms are abundant in the Arctic and Antarctic and in temperate and boreal coastal waters. Dinoflagellates are motile forms with two unequal flagellae. Some are armoured with elaborate cellulose plates. They are especially abundant in tropical and subtropical waters and in temperate and boreal water plankton in late summer and early autumn. Some forms (e.g. *Gonyaulax*) produce toxins and during bloom conditions may cause 'red tides' which occasionally result in mass mortalities of fish and invertebrates. Consumption of contaminated shellfish can cause a serious neurological disorder known as paralytic shellfish poisoning.

Other less important components of the phytoplankton include coccolithophores, blue-green prokaryotes and silicoflagellates. Nanophytoplankton consists of a heterogeneous mixture of especially small phytoplanktonic organisms; these are important primary producers in view of their exceptionally high reproductive rates.

Zooplankton

In contrast to the relatively limited taxonomic range of the phytoplankton the zooplankton includes representatives of virtually every animal phylum, either as adults (holoplankton) or as eggs and larvae (meroplankton). Copepod crustaceans are especially important members of the zooplankton. They are the *primary consumers* of phytoplankton and provide the vital link between the primary producers and secondary consumers (carnivores). Another group of crustaceans, the euphausiids, are abundant in many highly productive upwelling systems, and in the Antarctic they provide the principal food of the baleen whale. Comb jellies, arrow-worms, pelagic chordates, jellyfish and siphonophores are all important zooplanktonic predators.

Meroplankton

Approximately 70% of the world's benthic invertebrates produce planktonic (pelagic) larvae. This bewildering array of larval forms constitutes the meroplankton. Two major types of larvae, however, can be recognized. *Planktotrophic* larvae are produced in large numbers and spend prolonged periods feeding on phytoplankton or organic detritus; *lecithotrophic* larvae are short-lived and utilize yolk reserves during their development in the plankton.

Bacterioplankton

Pelagic bacteria feed on DOM released from living algae and from decaying organisms. The bacterioplankton is now identified as an exceedingly important source of primary production in pelagic water.

Adjacent bodies of neritic and oceanic water are frequently populated by different plankton communities due largely to different patterns of water circulation and sedimentation and to the differential effects of solar radiation. Shallow inshore waters are subject to greater fluctuations, especially in temperature and salinity. Accordingly, neritic species must be more physiologically tolerant than their oceanic counterparts. Neritic plankton contains numerous meroplanktonic forms, whereas oceanic plankton is more usually dominated by holoplanktonic species, though it may include a few particularly long-lived larvae (e.g. the phyllosoma larva of the lobster and the leptocephalus larva of the eel). For excellent accounts of marine plankton the reader should consult Raymont (1980; 1983).

Adaptations amongst planktonic organisms

Both phytoplankton and zooplankton must spend prolonged periods in the euphotic zone, the former to photosynthesize, the latter to feed on phytoplankton. Because planktonic organisms can swim only weakly, if at all, they must counteract gravitational forces which would otherwise cause them to sink. Sinking rate can be reduced in one of two ways.

(1) By reducing body density. Some species actively exclude heavy bivalent chemical ions (e.g. sulphate) and replace them with osmotically similar but lighter monovalent ions (e.g. chloride). Gas-filled floats and oils and fats may also serve to make planktonic organisms more buoyant.

(2) By increasing the surface area/volume ratio. Because frictional resistance is proportional to surface area, small organisms sink more slowly than larger organisms in view of their increased surface area. Small size also facilitates rapid multiplication and uptake of nutrients by phytoplankton cells. Areas where there is a premium on efficient nutrient uptake (e.g. open ocean), and on rapid multiplication to offset intense grazing pressure, therefore generally tend to support smaller phytoplanktonic species (see p. 32). Flattened shapes and elaborate spines and projections occur in many planktonic organisms. Such developments increase the surface of resistance and thus effectively slow the rate of sinking. Spiny projections can also be viewed as adaptations to increase the difficulty of capture and ingestion by grazers and predators.

Warm water is less viscous than cold water. Accordingly tropical species, and to a less marked extent summer populations at higher latitudes, tend to have more extreme modifications than species inhabiting cooler water.

Transparency is an important anti-predator adaptation amongst many pelagic organisms. The bright blue colour of many organisms living at the sea surface may serve to reduce predation from sea-birds and probably affords some protection against the damaging effects of UV radiation.

Light, nutrients and the seasonal cycle

Because photosynthesis is light-dependent, the phytoplankton is largely restricted to the euphotic zone. The amount of light reaching the sea surface which then actually enters the water column depends on the angle at which it strikes the surface, and this in turn is related to the maximal height of the sun above the horizon. Penetration is greatest at the equator. Moreover, light intensities in the tropics are optimal throughout the year, whereas at lower latitudes they show marked seasonal variation.

Within the water column, light intensity decreases exponentially with increasing depth as sunlight is rapidly absorbed or scattered by POM, planktonic cells or by the water molecules themselves. This attenuation of light is more pronounced in turbid coastal waters than in clearer oceanic water. Water, however, does not absorb all wavelengths of light equally; red and violet components are rapidly absorbed whereas blue and green light penetrate to much greater depths. Even so, the actual intensity of any given wavelength (measured as the extinction coefficient, i.e. the intensity at any given depth/surface intensity) decreases progressively with depth.

Photosynthesis increases logarithmically with light intensity to a maximal value when the system becomes light-saturated. Higher intensities may actually inhibit photosynthesis by bleaching photosynthetic pigments or arresting pigment production. A similar pattern emerges when the distribution of photosynthesis with water depth is considered (Figure 2.3). Very high light intensities at the sea surface may result in photoinhibition, whilst with increasing depth light intensities gradually decline to limiting values. Unlike photosynthesis, respiration does not vary appreciably with depth, and at some depth, therefore, the energy gained in photosynthesis exactly balances that lost by respiration. This *compensation depth* marks the lower boundary of the euphotic zone (where light intensity falls to 1% of its surface value). Above this level net production of plant material occurs, whilst below this level there is net loss of organic material as respiration exceeds photosynthesis. The light intensity corresponding to the compensation depth is termed the compensation light intensity. The compensation depth varies daily and seasonally as well as geographically

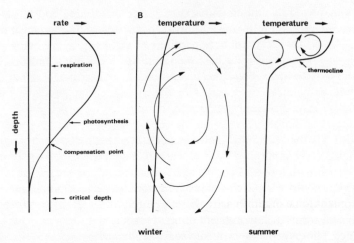

Figure 2.3 Relationship between (*A*) photosynthesis, respiration and depth; (*B*) wind-induced vertical mixing and temperature in inshore temperate–boreal waters during summer and winter.

according to the prevailing light conditions. In clear oceanic water it may extend down to 250 m whilst in very turbid coastal waters (e.g. estuaries) it may only be a few centimetres deep. However, different phytoplanktonic species have different optimal light intensities for maximal photosynthesis; this is probably important in phytoplankton succession (p. 27).

During winter in inshore temperate–boreal waters surface winds cause extensive vertical mixing of the water column. Under these conditions the phytoplankton is carried down below the compensation depth (Figure 2.3). When mixing is especially vigorous, the time spent in the euphotic zone in active photosynthesis may be insufficient to counteract the prolonged periods spent at greater depths. The *critical depth*, which is always below the compensation depth, occurs where the *total* production in the water column exactly balances the total consumption through respiration. The important factor here is the relative amount of time that the phytoplankton spends in the euphotic and aphotic zones, and this in turn depends on the degree of vertical mixing. When the mixing depth is less than the critical depth, photosynthesis exceeds respiration, resulting in net production of organic material.

In addition to dissolved carbon dioxide the phytoplankton also requires various nutrients for healthy growth. Of these the most important are nitrogen (as nitrate and ammonia) and phosphorus (as orthophosphate). Silicon is important to diatoms for frustule formation. Trace elements (e.g. iron, copper, vanadium) may be required in smaller amounts, and organic nutrients (e.g. vitamins) are also important in some cases. Nutrients are present in sea water in much smaller amounts than is carbon dioxide, and they must therefore be recycled to the autotrophs if production is to continue. Nutrient supply, especially of nitrogen, is probably the most common rate-limiting process in the sea (Ryther and Dunstan, 1971).

Work on the kinetics of nutrient uptake suggests that oceanic phytoplankton living in nutrient-poor waters is relatively more efficient at taking up nutrients at low concentrations. Species in nutrient-rich areas (e.g. inshore water) are less efficient but take up larger total amounts. Oceanic phytoplanktonic organisms are therefore probably competitively superior to coastal forms at low nutrient concentrations.

Since photosynthesis is a surface phenomenon, the euphotic zone frequently becomes temporarily depleted of nutrients, as these are used by the phytoplankton. Whilst some inorganic nutrients are regenerated and recycled in the euphotic zone, large quantities are ultimately incorporated into particulate detritus (e.g. as dead plankton and faecal material) which gradually sink to the sea bed. Most nutrient regeneration therefore occurs at greater depths, and there is a gradual draining of nutrients from surface waters. Exchange processes such as upwelling and turbulent mixing are therefore exceedingly important in restoring nutrients to the euphotic zone. Nutrients are also regenerated directly by consumers as excretory products and these are immediately available to phytoplankton and bacteria. Most nutrients, however, are regenerated by decomposition within the sediment (Chapter 4). Dissolved organic material is released (often in large amounts) by both algae and animal consumers and this too is recycled. Bacteria are the principal consumers of DOM in the water column, converting it into particulate organic matter (POM) which is then available to a variety of consumers such as protozoans and copepods. Bacteria are now known to be an important source of food in many marine systems.

Although much is now known about the size of various nutrient pools in the sea, we are still largely ignorant of the exchange rates which occur through any part of the ecosystem and it is the latter which ultimately govern photosynthetic production.

In temperate waters, variations in light and nutrient levels in the euphotic zone result in marked variations in the abundance of plankton (Figure 2.4). In winter the water column is more or less isothermal, and in the

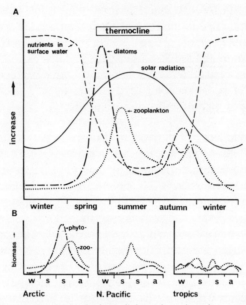

Figure 2.4   (*A*) Seasonal cycles in plankton, light and nutrients in temperate-boreal inshore waters. (*B*) Plankton cycles in three different geographical areas. (After Parsons *et al.*, 1984.)

absence of any marked density gradient (= pycnocline) with depth, surface winds cause extensive vertical mixing. Whilst nutrient levels are super-abundant, phytoplankton growth is effectively prevented by poor light conditions and by mixing of plant cells to greater depths. With the onset of spring, surface waters heat up and become less dense than the underlying layers. This impedes vertical mixing and the water column becomes thermally stratified. This stabilization of the water column is important since it effectively maintains the phytoplankton in the euphotic zone. Solar radiation increases, and although nutrient levels are initially high they soon decline as they fuel the developing spring diatom bloom. During summer, nutrient levels in the surface water become severely depleted and there is little replenishment from deeper water since vertical mixing is prevented by the pronounced thermocline. The volume of the phytoplankton is reduced by grazing zooplankton and by a shortage of nutrients; at the same time phytoplankton cells are continually sinking out of the euphotic zone. By autumn the surface water is cooling down and there is less light available for active photosynthesis. However, the breakdown of the thermocline and the re-establishment of vertical mixing leads to a rapid replenishment of nutrients in the surface water and this is often sufficient to stimulate a small, but short-lived increase in the phytoplankton.

In coastal waters wind and other forces (Chapter 1) generate turbulent mixing. In many nearshore situations, therefore, upwelling of nutrients may be an almost continuous process. This is one reason why coastal waters are sites of high productivity.

The zooplankton achieves its maximum abundance after the spring diatom bloom begins to wane; a slight increase may also follow the autumnal phytoplankton peak.

The seasonal pattern described above varies geographically (Figure 2.4). Tropical waters are well illuminated and thermally stratified throughout the year. Thus although light conditions are favourable for phytoplankton growth this is severely inhibited by the low nutrient content of the surface waters. No obvious relationship between phytoplankton and zooplankton can be discerned. In the Arctic a single large summer phytoplankton bloom when light conditions are favourable is followed by a single zooplankton peak. Nutrients are not limiting and the water column is never strongly stratified. Phytoplankton growth is here primarily limited by light. In the North Pacific, intensive grazing by the zooplankton effectively prevents the phytoplankton from reaching bloom conditions.

Apart from seasonal changes in phytoplankton abundance, marked changes in species composition also occur. This change in species composition over the course of the year is termed seasonal *succession* (see Margalef, 1962). Although most pronounced in areas with marked changes in temperature, certain successional changes also occur in polar and tropical seas. The precise causes of succession are still unknown, though biological conditioning of the water by the organisms in it could be important. Certain species for instance may depend on metabolites (e.g. vitamins) produced by species earlier in the successional sequence. Dinoflagellates typically require more nutrients which they are unable to synthesize themselves than do diatoms, perhaps explaining their generally later appearance in the plankton. Some organisms produce toxic allelochemicals which inhibit growth in other species. Changes in temperature, light and nutrient levels may also be important determinants of seasonal succession.

Zooplankton communities in temperate–boreal waters are initially dominated by calanoid copepods which graze on the abundant diatoms. Meroplanktonic forms appear in late spring and early summer. Predatory species become abundant in summer and feed extensively on the herbivores.

Grazing

The zooplankton generally appears in abundance after the major peaks in the phytoplankton (Figure 2.4) suggesting that herbivorous grazing contributes to the decline of the phytoplankton bloom. In some areas, however, nutrient limitation is also important and the precipitous decline in the spring diatom bloom is therefore probably due to the combined effects of grazing and depletion of nutrients as these become locked away with the spring stabilization of the water column. It should be noted, however, that in the Antarctic, where nutrients are not limiting, the steep drop in phytoplankton numbers is still evident. The most important group of planktonic grazers is the copepods (accounting for 70–90% of the herbivore biomass). Moreover, in many areas only a few species of copepods are numerically abundant.

Copepods are selective grazers (e.g. Conover, 1978) capable of drastically reducing phytoplankton populations in patches of ocean. Moderate levels of grazing, however, probably promote phytoplankton production since plant cells are broken up, thus recycling nutrients back into the water. Animal excretions also stimulate phytoplankton growth. Without herbivores phytoplankton production may eventually cease, especially in areas where nutrients are severely limiting.

Copepods usually graze the most abundant size classes of the phytoplankton. Such selective grazing therefore favours the rarer species and less abundant size classes, thus influencing species composition and preventing any one species from becoming competitively dominant. Because feeding is generally suppressed when phytoplankton density falls below a threshold value, grazing to extinction is effectively prevented. This low-density 'refuge' enables the phytoplankton to increase in abundance whenever grazing pressure is subsequently reduced. Phytoplankton blooms start (1) because of the time lag in zooplankton reproduction and (2) because phytoplanktonic organisms have shorter generation times than those of copepods. Sometimes, however (e.g. North Pacific) herbivores appear to be able to 'predict' the onset of bloom conditions and breed in advance using accumulated food reserves.

The effectiveness of grazing varies not only seasonally but also spatially. In coastal waters of considerable environmental variability, prolonged periods may occur when the phytoplankton is not effectively controlled by grazing. This contrasts with more open water conditions where grazing appears to be more important throughout the year. In the tropics the phytoplankton is limited to low levels at all times and is continually consumed by copepods.

Spatial patterns

Patchiness

The distribution of plankton in the euphotic zone is often very patchy (see Steele, 1978). Patches range from a few metres to over 100 km in width. Patchiness is more pronounced in coastal waters where conditions (e.g. salinity. temperature, grazing) are generally more variable. Open ocean plankton, by contrast, tends to be rather more uniformly distributed. Various processes promote patchiness.

(1) Large-scale patterns of current circulation

(2) Turbulence. Any localized turbulent mixing which raises nutrient concentration will enhance phytoplankton production. Coastal upwelling for instance is an important cause of plankton patchiness. Steady winds blowing persistently over the sea surface create shallow convection cells resulting in small-scale divergences (at upwellings) and convergences (at downwellings). Plankton becomes concentrated in these Langmuir circulations (Figure 2.5) on the basis of their buoyancy (Smayda, 1970).

(3) Grazing. Aggregations of herbivorous zooplanktonic organisms may effectively graze an area of dense phytoplankton to virtual extinction before moving elsewhere and thus leaving the original patch to recover. This frequently results in an inverse relationship in the densities of phytoplankton and zooplankton. Some phytoplankton patches, however, produce toxic metabolites which are repellent to grazers.

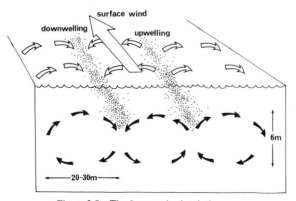

Figure 2.5 The Langmuir circulation.

(4) Reproduction. When phytoplankton cells reproduce, the daughter cells tend to remain close together, thus emphasizing any original unevenness in horizontal distribution. Social behaviour within zooplanktonic populations and interspecies attractions or repulsions are additional factors promoting patchiness.

Plankton is also patchily distributed within the water column. Phytoplankton occurs at depths with optimal light intensities for photosynthesis. Photosynthesis is maximal in the subsurface layers and one striking feature of vertical patchiness is the subsurface chlorophyll maximum. The precise level at which this occurs in temperate–boreal waters varies both seasonally and as a result of the self-shading effects created by phytoplankton cells themselves. Plankton also concentrates at boundary layers (e.g. where water bodies of different temperature and salinity meet). Differential sinking rates between species also influence the vertical distribution of the plankton.

Vertical migration

Many zooplanktonic species undertake extensive (> 100 m) diurnal vertical migrations. Although this complex behaviour pattern varies extensively both within and between species there is, nevertheless, a remarkable consistency in the general pattern (Figure 2.6)

During daylight the zooplankton occurs deep in the water column. At dusk organisms ascend to the surface to feed on the phytoplankton. During total darkness they disperse, but at dawn they reaggregate at the surface

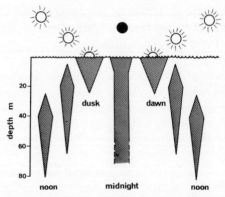

Figure 2.6   Daily vertical migration pattern of many zooplanktonic organisms.

before descending to their daytime depth. Larger species can swim at speeds of up to $200\,m\,h^{-1}$. Movement may be regulated by an internal biological clock, since some species start their descent before dawn whilst others maintain their activity patterns even when experimentally kept in total darkness.

Diurnal variations in light intensity appear to provide the proximate stimulus for migration but the adaptive significance of this remarkable phenomenon is still far from clear. Several explanations have been proposed.

(1) Avoidance of high light intensities. Migrations often appear to be too extensive for this to be a likely explanation. Moreover, some deep-water species also migrate.

(2) Avoidance of visually-hunting surface predators. Many zooplanktonic species bioluminesce during their migrations (a factor that is hardly likely to conceal them from predators) and many potential predators are themselves migratory.

(3) Improved dispersion. Migration may enable weakly-swimming species to change their horizontal position, since waters at different depths move in different directions and at different speeds. This would also effectively prevent the zooplankton from outgrazing local phytoplankton stocks and might explain the patchy distribution noted earlier. Discontinuous grazing enhances phytoplankton production and thus increases the supply of food to the zooplankton.

(4) Lower metabolic cost. It may be energetically profitable for the zooplankton to spend part of the time in deeper cooler water where metabolic rate is reduced. However, the energy thus saved must be set against the amount of energy expended migrating. Evidence suggests that the metabolic cost of migrating is probably minimal. It is noteworthy that vertical migration is generally least strongly developed in relatively isothermal seas where the temperature differential between surface and deeper waters is minimal. Fecundity of some species also seems to be enhanced by part-time residence in deeper water. It has been suggested therefore that a diversity of benefits is probably conferred on species exhibiting vertical migration (e.g. Longhurst in Cushing and Walsh, 1976).

Diurnal changes in distribution are also exhibited by fish and cephalopods. Echo-soundings have shown that 'deep scattering layers' first noted during World War II, comprise dense aggregations of migrating planktonic and nektonic animals. Some phytoplankton species are also known to undertake limited vertical migrations. In shallow waters many benthic organisms leave the sea-bed at night and swim in the surface waters. Apart from the diurnal vertical migrations noted above, some species migrate seasonally, descending in the autumn to overwinter at greater depths. These migrations too may have evolved for conserving energy.

Trophic interrelationships

The description of the trophic structure of a community has been based upon the concept of the food chain. Organisms have conventionally been grouped into categories or trophic levels (producers, consumers, decomposers) according to their position within the food chain. Most autotrophs (primary producers) fix carbon through photosynthesis. These then provide food either directly or indirectly for a variety of heterotrophs (consumers) which occupy many trophic levels. Primary consumers feed directly on producers, secondary consumers are carnivores feeding on the primary consumers, tertiary consumers feed on the secondary consumers and so on. Decomposers (bacteria, fungi) break down organic detritus, releasing simple molecules which are re-used by the autotrophs. Although a convenient abstraction, the trophic level concept is a gross oversimplification. Species change their feeding habits during their life cycles and are rarely, if ever, confined to a single trophic level. In particular, detritus feeders utilize material derived from several trophic levels.

Three basic types of food chains associated with oceanic (5 trophic levels) shelf (3 trophic levels) and upwelling regions (1–2 trophic levels) can be recognized (Ryther, 1969). In part, the different lengths of these chains are related to the relative size differences between the dominant phytoplankton species and the nektonic consumers. In nutrient-poor water (e.g. open ocean) phytoplanktonic species are small in size (nanophytoplankton and bacterioplankton). Consequently these are grazed only by small zooplanktonic organisms and energy must be concentrated through several further trophic levels before larger members of the nekton (e.g. fish) enter the chain. In nutrient-rich areas, on the other hand, dense populations of larger phytoplanktonic species (e.g. chain diatoms) can be consumed directly by large zooplanktonic species or even by planktivorous fish such as anchovies. Since a considerable amount of energy ($< 90\%$) is always lost between successive trophic levels (e.g. as respiratory heat), shorter chains typical of nutrient-rich coastal and upwelling areas increase the yield of the top trophic levels with consequent implications to the commercial fisheries (see Landry, 1977).

Environmental stability may also contribute to food-chain length. It has been argued that complex food chains tend to be inherently unstable; shorter chains should therefore be favoured in the environmentally unstable conditions more typically associated with turbulent coastal environments.

Productivity

Life in the sea depends largely on the ability of photosynthetic plants to utilize solar energy to synthesize energy-rich organic molecules from simpler inorganic molecules. This process is known as *primary production*. Although macrophytic plants (algae, mangroves, saltmarsh plants, sea-grasses) contribute substantially to inshore production, most primary production in the sea is attributable to surface phytoplankton. Chemosynthetic bacteria are important primary producers below the euphotic zone. Production, or rate of carbon fixation (generally expressed as $g\,C\,m^{-2}\,y^{-1}$) is quite distinct from *standing crop* or biomass which describes the amount of organic material present in any given volume of water at any given time. A population may have a low standing crop not because of low productivity but simply because of high grazing pressure. Similarly, a high standing crop need not necessarily indicate high productivity. Secondary and tertiary production refer to the production by herbivores and first-rank carnivores respectively.

Primary production is usually measured indirectly as the amount of oxygen evolved during photosynthesis, or directly using labelled carbon (as ^{14}C in HCO_3^-). However, because phytoplankton cells exude much of their photosynthate, the radioactive carbon method generally gives a lower estimate of photosynthesis than the oxygen evolution technique. The total amount of energy fixed during photosynthesis is termed *gross primary production*. Some of this production is utilized by the plants themselves for respiration and a lesser amount ($=$ *net primary production*) is thus available to the consumers. Annual net production provides a convenient way of comparing various ecosystems.

Several attempts have been made to estimate primary productivity in the sea on a global scale, though values vary (e.g. $25{-}200\,g\,C\,m^{-2}\,y^{-1}$) according to the analytical techniques used, and valid comparisons are therefore difficult. A value of $50\,g\,C\,m^{-2}\,y^{-1}$ is probably a realistic world average though this is subject to broad geographic variation. Primary production is generally low in open ocean communities. Nutrient upwelling areas are especially productive ($300\,g\,C\,m^{-2}\,y^{-1}$) and some of the

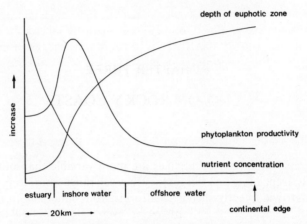

Figure 2.7 Variation in phytoplankton productivity, nutrient concentration and depth of the euphotic zone across the continental shelf. (After Haines, in Livingston, 1979.)

world's largest fisheries (e.g. for Peruvian anchovy) are located in these regions. Frontal systems, where water masses of different properties meet at sharp boundaries, are also areas of high biological productivity. By virtue of their turbulent nature, shallow coastal waters are rarely nutrient-limited, and given a suitable light regime high levels of production can occur throughout much of the year. In estuaries, the advantage of high nutrient input (from terrestrial origin) is offset by high turbidity which severely reduces the depth of the euphotic zone (Figure 2.7). Differences between inshore and offshore water are especially pronounced in the tropics, presumably reflecting the increased levels of nutrients in the immediate coastal zone.

Whilst primary production has been extensively studied, much less is known about zooplanktonic production. In general, however, secondary production parallels primary production, being high $(27.5\,g\,C\,m^{-2}\,y^{-1})$ in coastal and upwelling regions, and much lower $(4.5\,g\,C\,m^{-2}\,y^{-1})$ in tropical oceanic waters. However, exceedingly high levels of primary production can sometimes lead to anoxic conditions similar to those characteristic of eutrophic lakes, and these can drastically reduce levels of secondary production. Moreover, particulate detritus and bacteria utilizing DOM may provide important alternative energy sources for many pelagic consumers. Until these pathways have been adequately quantified, any estimates of secondary production must remain suspect.

CHAPTER THREE

LIFE ON ROCKY COASTS

Topographically, rocky shorelines are more variable than other coastal habitats. Depending on local geology they can range from steep, overhanging cliffs to wide, gently-shelving platforms, from smooth uniform slopes to highly dissected, irregular masses or even extensive boulder beaches (Lewis, 1977). Many open coasts are continuously exposed to oceanic swell and extreme wave action, whilst deeply indented coastlines may be perpetually calm. Salinity, temperature and turbidity are also subject to wide variations.

The intertidal zone has been especially well studied. This in part reflects the sessile or sluggish nature of the common species whose populations are therefore easily estimated, and in part the relative ease with which intertidal communities can be experimentally manipulated (see Paine, 1977). Most early investigations were entirely descriptive, but more recently emphasis has shifted towards an understanding of the dynamic relationships between the organisms and their physical and biological environment. Some representative rocky-shore organisms are illustrated in Figure 3.1.

Life habits and adaptations

Organisms living on open rock surfaces (epifaunal species) are more exposed to the rigours of the physical environment than infaunal species inhabiting soft sediments (Chapter 4). Substratum therefore is arguably the most important factor determining the distribution and adaptive characteristics of benthic organisms. Rock presents a stable surface to which organisms must attach or into which they must bore. Surface texture depends on rock type and weathering and is important in determining attachment success; mobile species need to move over the surface but must also resist dislodgement by waves and currents.

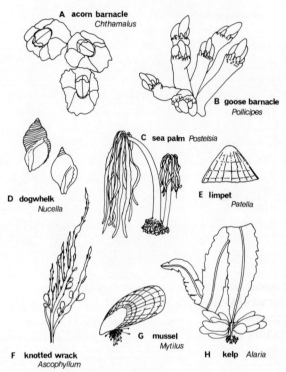

Figure 3.1 Some representative rocky-shore organisms (redrawn from various sources). Not to scale.

Water movement

Maintaining position in the face of strong water movement frequently involves one or more of the following adaptive features: (1) strong *attachment devices*, e.g. algal holdfasts, cementation (oysters, barnacles), byssus threads (mussels), tube-feet (urchins, starfish), the adhesive foot of gastropods and the modified sucker-like fins of many pool fish; (2) *boring* into the rock surface; (3) *changes in orientation* to minimize shear stress; (4) use of *crevices*; (5) formation of dense *aggregations* to expose a smaller surface area; (6) increased *flexibility*, e.g. algal stipes; (7) *irregular surface contours* to reduce turbulence and minimize drag, e.g. the ridged or crinkly fronds of many kelps.

Desiccation, temperature and light

Intertidal organisms, especially those from the high shore, must be resistant to extreme physical conditions. By virtue of tightly-fitting opercular plates and a non-porous shell the high-shore barnacle *Chthamalus montagui* (previously *C. stellatus*) is more resistant to desiccation than the other commonly occurring intertidal barnacle *Semibalanus* (= *Balanus*) *balanoides*. Snails protect themselves from excessive water loss by means of a horny or calcareous operculum. The shell aperture of neritid gastropods is more restricted than in limpets (*Patella*); neritids are therefore better adapted to withstand desiccation and usually replace limpets on tropical shores. Some limpets, however, reduce water loss by returning to a 'homescar' where their shells neatly fit the contours of the rock and form an effective seal. Many high-shore algae (e.g. *Pelvetia*) can lose substantial amounts of water apparently with little ill effect. In some species the water appears to be lost from mucus-like material which surrounds and protects the cells.

Many tropical molluscs have pale, heat-reflecting shells. Ridges, spines and other forms of shell sculpturing serve to increase surface area and may act as effective radiators. When subjected to heat shock, high-shore species of *Nerita* exhibit greater cell thermostability than species from the lower intertidal zone. Amongst British snails, heat coma ranges from 28°C in the dogwhelk (*Nucella* (= *Thais*) *lapillus*) to 40°C in *Littorina neritoides*, a snail that inhabits the high splash zone. High summer temperatures frequently restrict many delicate algae (e.g. *Ulva, Enteromorpha*) to pools or damp crevices. Low temperatures can be equally damaging, though many intertidal organisms (e.g. *Mytilus, Fucus*) can survive extended periods of sub-zero (− 20°C) temperatures. The effects of freezing are similar to those caused by drying in that body fluids become increasingly concentrated. Mobile organisms (e.g. many crabs and snails) avoid extreme temperatures by migrating with the tide or by seeking refuge in crevices and weed-beds during low-tide exposure.

High light intensity can be damaging to plants. In some algae (e.g. *Ulva, Codium*) photosynthetic pigment is housed in chloroplasts which retreat from the cell surface at high light intensity. Other species have masking compounds (e.g. polyphenol granules) that protect the chlorophyll molecules from photolysis. The bright pigmentation of many high-shore lichens may serve to screen out excess light.

Feeding strategies

Many of the organisms which occur on rocky coasts are sessile. The advantages of this on wave-swept shores are obvious, but it does limit the methods of food acquisition. Not surprisingly, suspension feeders are particularly prominent. These utilize various structures including gills (bivalves), setae (barnacles), tentacular crowns (tubeworms, bryozoans) and pharyngeal baskets (ascidians) to collect particulate material. Some species also extract DOM from the large volumes of water which pass over the feeding structures.

Sea urchins, chitons and gastropods are the principal herbivores. Urchins use a complex series of teeth (Aristotle's lantern), chitons and gastropods a file-like radula to scrape the rock. Common predators on herbivores include crabs, whelks, starfish and birds. Some of these have generalized diets which include many different prey items, others have highly specialized food requirements. Predators often select an optimal prey size to maximize their energy intake. Some prey species, however, occasionally grow too large for their predators to consume them. Such *size refuges* effectively release these prey from further predation.

Most marine species feed only when submersed, and reduced feeding time may therefore limit the extent to which some species can penetrate the intertidal zone. Predators in particular are inhibited by prolonged tidal emersion. Consequently for many prey species the high shore, although a suboptimal habitat, constitutes an effective *spatial refuge* in which survival may be greatly enhanced. Refuges from predation are probably essential for the long-term coexistence of predator–prey populations.

Various antipredator devices have also evolved. Structural adaptations include spines (urchins), thickened shells with narrow openings (many gastropods) and calcified tissues (crustose algae). Mobile prey may have escape behaviours and often respond to water-soluble chemicals exuded by the predator. Chemical defences such as acidic secretions and toxins (e.g. phenols, tannins) are also widespread. Many of these adaptations are more prevalent in the tropics where predation is frequently more intense.

Life cycles

Most benthic invertebrates produce planktonic larvae. Planktotrophic larvae are produced in large numbers and spend prolonged periods (weeks to months) in the plankton; lecithotrophic larvae are short-lived (hours to days) and because they rely on stored yolk for their nutrition they are produced in smaller quantities. Although planktonic larvae provide an effective means of dispersal (especially in planktotrophic types) they are subject to high mortality, especially from predation. In some species the planktonic stage is suppressed and development is direct; juvenile dog-whelks for instance emerge from tough egg capsules cemented to the rock, and the anemone *Actinia equina* broods its young within the gastric cavity. The benefits of widespread dispersal are therefore lost to these species but the risks associated with producing vulnerable larvae are also correspond-ingly diminished. Asexual proliferation permits the rapid spread of advantageous genotypes but is rarely the exclusive means of reproduction. Sexual reproduction promotes genetic variability and the potential to adapt to new or changing environments.

The life cycles of marine algae are extremely complex and may involve two or even three morphologically distinct phases (heteromorphism). In some species an asexual (diploid) sporophyte stage alternates with a sexual (haploid) gametophyte stage, but this basic pattern is subject to extensive variation. Although enormous numbers of spores are produced, dispersal distances may be quite short. The sea-palm (*Postelsia palmaeformis*) grows in severely wave-swept conditions along the Pacific coast of North America. Spores released at low tide follow the corrugations along the fronds and drip on to the underlying rock close to the parent plant. Long-distance dispersal occurs only when fertile parts of the plant break free and are carried away by the waves. *Bifurcaria brassicaeformis* also lives in very exposed areas where spore settlement is difficult. Here the spores begin to grow whilst still attached to the plant by means of a gelatinous stalk; only later do they extend onto the rock surface. *Porphyra* has a particularly interesting cycle involving a perennial filamentous stage (conchocelis) which bores into mollusc shells. This subtidal stage provides an effective refuge from herbivores; periodically it produces conchospores which colonize the intertidal whenever conditions are suitable.

Zonation—a descriptive framework

Biologically the intertidal zone is overwhelmingly a marine province. It is inhabited principally by marine organisms and a few terrestrial species tolerant of short periods of tidal submersion. Because marine organisms tolerate aerial exposure to varying degrees, species progressively replace each other along the marine–terrestrial gradient. This results in *vertical*

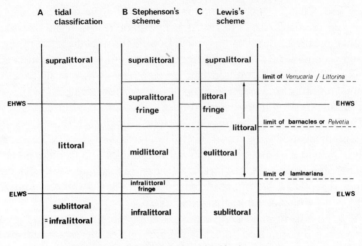

Figure 3.2   Schemes used to classify shore zones.

zonation where each species is typically most abundant within a particular zone where conditions favour its survival; above and below this optimal zone abundance declines either as the physical environment becomes less suitable or because the species interacts less favourably with other species. Wherever sessile organisms occur in abundance to the limits of their vertical distribution, especially on steep rock faces, they appear as conspicuous lines (e.g. barnacle line.). Mobile species can migrate into favourable microhabitats and exhibit less well-defined zonal boundaries.

All early attempts to classify these vertical zones were based on tidal levels. Individual species were placed within a broad framework defined on physical (= tidal) criteria (Figure 3.2). In this scheme the terms 'littoral' and 'intertidal' are synonymous. However, although zonation is indeed related to tides, it is also influenced by many other factors (p. 42) which vary from one shore to another. As noted by Lewis (1964) it is this general lack of coincidence between observed distributions and tidal levels that has caused some of the basic terminological problems in shore ecology. Stephenson's system of classification was a major advance for it uses biological rather than physical criteria to define the major zones (Figure 3.2). Although species composition varies geographically the *universal* occurrence of these biological zones provides a convenient framework within which local features can be accommodated without obscuring those underlying features by which *all* rocky shores can be compared. This scheme was subsequently modified by Lewis (1964) who recognized three biologically defined zones—the littoral fringe, the eulittoral and the sublittoral (Figure 3.2). The term 'supralittoral' is reserved for the lowest belt of terrestrial vegetation typically dominated by orange and grey-green lichens. Here the littoral zone is distinct from the intertidal (= shore) which is a physical rather than a biologically defined entity. The significance of using biological criteria to define zones will become evident when we consider the determinants of vertical zonation.

The littoral fringe is a relatively arid zone transitional between land and sea. Relatively few species occupy this zone which is dominated by small littorinid snails, black crustose lichens (*Verrucaria*) and small blue-green prokaryotes (Cyanophyta). The eulittoral is occupied mainly by barnacles, though mussels and other sessile forms (e.g. oysters, zoanthids, vermetids) may also be abundant. A narrow band of red algae is often present towards the lower limits of this zone. On many North Atlantic coasts large brown fucoid algae predominate in regions sheltered from wave action. These too exhibit a distinct pattern of zonation (Figure 3.3). The sublittoral is uncovered only during spring tides. It is a particularly rich and variable zone which in cold temperate, boreal and antiboreal regions (Figure 4.6) is

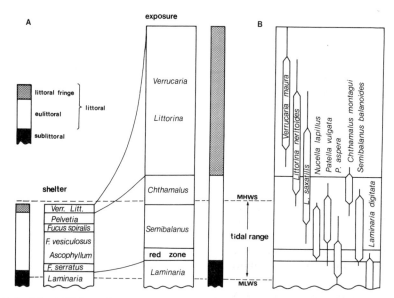

Figure 3.3 (*A*) Effects of wave exposure on zonation patterns along a typical North Atlantic shore. (*B*) Distribution of some common rocky shore organisms within the framework provided by the major biological zones.

dominated by kelps. In more tropical regions it contains rich growths of coral (Chapter 6) and in warm temperate waters often supports ascidians (e.g. *Pyura*) or turfs of red algae. For details of rocky shores worldwide see Stephenson and Stephenson (1972).

Factors influencing zonation

Physical factors

Zoned distributions are primarily created by the varying responses of marine organisms to environmental stresses which are themselves related to the tidal environment. At high tide, organisms are subjected to a relatively uniform temperature, there is no risk of desiccation and oxygen and food are readily available. At low tide, intertidal organisms are exposed to the full rigours of the physical environment. However, as already noted, there is generally no exact coincidence between zones and tidal levels because zonation patterns are *modified* by many factors other than tides. Amongst these, wave action is especially important. The effect of waves is to raise the upper limits of marine organisms (Figure 3.3) since splash and spray keep the rock surface permanently wet above the level normally reached by the tide. This uplifting effect, which is often evident even on a very localized scale, for example on the exposed and sheltered sides of headlands and breakwaters, can result in much of the eulittoral zone extending well above theoretical high-water mark; even kelps can penetrate beyond mean tide level on particularly waveswept coasts. However, it is the littoral fringe which experiences the greatest uplift; in shelter this is a very narrow zone completely contained within the range of tides but on very exposed coasts it may be several times wider (vertically) than the eulittoral zone (Figure 3.3). The width of this zone is therefore a valuable indicator of the degree of wave exposure. Such observations clearly emphasize the importance of using biological rather than tidal criteria to define the major zones. Factors other than wave action can also modify zonation patterns. These include topography (steepness), nature of the substratum, climate, aspect (i.e. the degree of shading) and the time of day at which low-water spring tides recur. In general, the effect of these factors is to raise the upper zonal limits wherever conditions are locally cooler and damper. Since zonation can therefore be modified by various factors operating on a local scale, the once widely held view that the limits of some species coincide with certain recurring critical boundaries at which there are abrupt changes in the frequency and duration of periods of continuous submersion and emersion (e.g. near the neap tide levels) must be highly questionable. The concept of 'critical tide levels' is reviewed by Underwood (1978).

Biological factors

Competition for limited resources (mainly space on rocky coasts), predation and grazing are also important determinants of zonation. Whilst the distribution of some organisms is centred mainly in the high shore there is no conclusive evidence to suggest that their extension downshore is prevented by physical factors. Indeed, when experimentally transplanted to the lower shore many of these organisms grow much faster. Their occurrence in what for marine species is clearly a more stressful, marginal habitat is determined largely by interactions with other species. This has been repeatedly demonstrated by controlled field experiments. The following selected examples serve to illustrate how physical and biological factors, and their complex interactions, influence the zonation of some of the principal inhabitants of rocky coasts.

Wherever the barnacles *Chthamalus montagui* and *Semibalanus balanoides* co-occur on British shores, *Chthamalus* always occurs higher in the eulittoral zone than *Semibalanus* (Figure 3.4). Although *Chthamalus* recruits throughout much of the eulittoral its subsequent disappearance at the lower levels is due to competition from the more rapidly growing, heavier-shelled *Semibalanus* which can overgrow, undercut or laterally

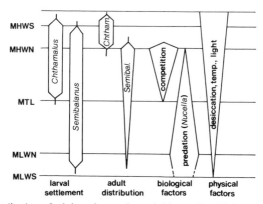

Figure 3.4 Distribution of adult and recently settled barnacles, *Chthamalus montagui* and *Semibalanus balanoides*. The relative effects of biological and physical factors are also shown. (After Connell, 1961.)

crush young *Chthamalus*. However, *Chthamalus* is more resistant to desiccation (p. 37) and can therefore extend further up the shore than *Semibalanus*. In the zone where the two species overlap, *Chthamalus* grows more slowly whenever its major competitor is present. On shores where *Semibalanus* is absent or artificially removed, *Chthamalus* extends downshore. However, desiccation effectively prevents the upshore extension of *Semibalanus* even when *Chthamalus* is absent. The lower limits of *Semibalanus* in the North Atlantic and *Balanus glandula* on the west coast of North America are set either by competition with mussels, which are more effective at sequestering space, or by dogwhelk predation.

On wave-swept shores along the Pacific coast of North America the mussel *Mytilus edulis* regularly occurs as a narrow band immediately above the zone occupied by its congener *M. californianus* (Suchanek, 1981). Selective predation by the dogwhelk *Nucella* (= *Thais*) *emarginata*, and competition from the heavier-shelled *M. californianus* effectively exclude *M. edulis* from lower shore levels. However, the ability to colonize quickly and grow rapidly to reproductive size (i.e. fugitive characteristics) allow *M. edulis* temporarily to colonize open patches of rock wherever its competitor is removed by waves or predators. Desiccation and heat stress clearly determine the upper zonal boundaries since large numbers of moribund mussels occur, particularly on hot summer days. The lower limits of *M. californianus* are set by the predatory starfish, *Pisaster ochraceus* (Paine, 1984). These results therefore broadly parallel those outlined above for barnacles.

The interaction between physical and biological factors has similar effects on the zonation of macroalgae. On the eastern seaboard of North America the mid-littoral zone is dominated by *Fucus vesiculosus*, whilst *Chondrus crispus*, a red alga, predominates at lower levels (Lubchenco, 1980). When *Chondrus* is experimentally removed, *Fucus* extends downshore provided that periwinkle snails are prevented from consuming the *Fucus* sporelings. Interspecific competition therefore normally restricts *Fucus* to the upper zone whereas desiccation prevents the upshore extension of *Chondrus*. The upper limits of *Pelvetia*, *Fucus spiralis* and *Ascophyllum* on sheltered shores in Britain are periodically pruned back by harsh environmental conditions; these species clearly reach their physiological limits on the shore. This was confirmed by laboratory experiments which showed that the ability to tolerate desiccation and then resume growth when re-submerged was greatest in *Pelvetia*, the highest-occurring fucoid, and was progressively less in species from lower tidal levels. The lower limits of *Pelvetia* and *F. spiralis* are probably set by interspecific competition since plants transplanted downshore grow well (Schonbeck and Norton, 1980).

The tidal restriction of most marine predators (p. 38) often results in distinct *predation lines* below which established prey populations are rare. On the east coast of England predatory starfish (*Asterias rubens*) and dogwhelks (*Nucella lapillus*) effectively remove *Mytilus edulis* from much of the lower shore (Seed, 1969). Predation in the upper shore, by contrast, is minimal and mussels growing there may therefore be exceptionally long-lived (> 20 y). On the Washington coast *Mytilus californianus* extends further downshore wherever its major predator (*Pisaster*) is experimentally removed. Similar downward shifts in the distribution of the mussels *Perna canaliculus* (in New Zealand) and *Perumytilus purpuratus* (in Chile) have been observed following removal of major predatory starfish, *Stichaster* and *Heliaster* respectively (Paine *et al.*, 1985). It is perhaps significant that those species which are known to be regulated by predation (or grazing) are predominantly sessile (e.g. barnacles, mussels, macroalgae). Mobile species (which probably compete mainly for food rather than for space) are perhaps less likely to be predator-limited.

Thus whilst the upper zonal limits of many intertidal species, particularly those from the high shore, are set ultimately by the extreme physical environment, many mid- to low-shore organisms seem to live well within their physiological limits. The upper limits of these organisms appear to be determined mainly by interactions with other species, many of which are better adapted to aerial exposure. However, with few exceptions, lower zonational boundaries are fixed by ecological rather than physiological factors.

Settlement and behaviour

The larvae of most benthic invertebrates are highly selective in their choice of habitat and actively search out suitable surfaces on which to settle. Some will even delay attachment until a suitable substratum is encountered. For sessile organisms it is particularly important that prospecting larvae locate and select a habitat compatible with adult survival since once attached they are unable to move elsewhere should conditions subsequently prove to be unsuitable. Various chemical and tactile cues elicit attachment (see e.g. Crisp, 1984). These include surface texture (many larvae prefer pitted or creviced surfaces) and inclination, biotic factors such as bacterial films, and either the presence of adults of the same species (gregariousness) or some indicator species with which the species predictably co-occurs. Barnacles settle only in a suitable current regime, an adaptation which prevents them from attaching in areas where sediment is likely to accumulate and smother them.

Settlement responses clearly require some form of recognition. Barnacle cyprids respond to quinone-tanned proteins in the adult exoskeleton, certain spirorbids and bryozoans to specific algal exudates. Settlement can often be induced on inert panels filmed with appropriate extracts. Chemosensory organs are therefore generally well developed in most marine larvae. Gregarious settlement can be detrimental in that it may lead to overcrowding and increased intraspecific competition. However, many gregarious organisms (e.g. barnacles, spirorbids) space themselves out from conspecifics allowing adequate space for firm attachment and early growth but denying space to potential competitors. Mussels spend a primary attachment phase on filamentous algae thus reducing intraspecific competition at a time when they are small enough to enter adult feeding currents. Only when they are somewhat larger do they migrate on to the established mussel bed.

In mobile species larval selectivity is largely replaced by behavioural responses in the adults. Territoriality and homing behaviour (e.g. limpets) and simple responses to directional stimuli such as light and gravity (e.g. chitons, littorinids) are adaptations which keep these organisms in their optimal zone.

Geographical and local distribution, patchiness

Given suitable substratum and favourable currents for larval dispersal, latitudinal gradients in sea temperature are the principal determinants of *geographical* distribution for most marine organisms. Air temperature, which has a much wider range than sea temperature, is also important for intertidal species. The surface sea-temperature gradient along the east coast of North America is greatly compressed (ranging from $< 5°$ to $> 25°C$ within a mere 20° of latitude) and the distribution of many intertidal species is therefore well correlated with critical thermal boundaries. On the west coast the gradient is less pronounced and other factors (e.g. suitable microhabitats, predation) become more important in determining distribution patterns. In Britain the distribution of many rocky-shore organisms corresponds closely with sea temperature (Lewis, 1964) and some species are known to advance or retreat with long-term shifts in climate. The precise manner in which temperature influences distribution (e.g. via its effects on adult survival, competitive ability or reproductive success) is less well documented and the underlying physiological mechanisms are probably extremely complex.

Man can influence distribution by introducing species, either accidentally or for commercial exploitation. The Australasian barnacle, *Elminius modestus*, is one such immigrant species and this now thrives on sheltered rocky shores throughout Britain, often at the expense of native barnacles. Construction of major waterways (e.g. the Suez and Panama canals) opens up new dispersal routes (see Chapter 11).

Present-day distributions, however, are determined not only by various physical and biological factors operating on an ecological time scale and having an immediate impact on the biota, but also by historical processes such as speciation, colonization and extinction, acting over evolutionary time.

Many of those factors which influence zonation also make some shores suitable for certain species but not for others. Wave action is an especially important factor controlling *local* distribution. Delicate species with poor powers of adhesion, and many long-fronded algae, are unable to withstand turbulent conditions and consequently they are largely restricted to quieter water. Other species, however (e.g. the kelp *Alaria esculenta* and the winkle *Littorina neritoides*) are typical of wave-beaten coasts and therefore can often be used as *indicators* of exposure. Waves also affect distribution indirectly by their effect on the physico-chemical environment (e.g. oxygenation, turbidity). Many species occur throughout the exposure gradient but are more abundant on exposed shores probably because of the increased supply of larvae (or spores). However, reduced intraspecific competition and favourable feeding conditions, especially under a damp canopy of macroalgae, often enable those larvae which do settle in shelter to grow to a much larger size.

Biological interactions also influence local distributions. The sweeping fronds of large algae, for example, may prevent successful larval attachment. In western North America *Mytilus edulis* is virtually excluded from the most wave-swept shores (or confined to a high-shore refuge, p. 44) through competition from *M. californianus* (Harger, 1972). The latter species grows to a large size and has a more robust shell and stronger byssal attachment than *M. edulis*, adaptations clearly suited for life on exposed shores. The ability to crawl out of accumulated sediment, however, gives *M. edulis* a competitive edge in shelter. In regions of intermediate exposure these species coexist throughout much of the intertidal region because of the heterogeneous nature of the environment, the balance of advantages shifting as conditions favour first one species then the other. By using cages to exclude or include predators, Ebling *et al.* (1964) demonstrated that the absence of mussels from areas of moderate wave exposure in south-west Ireland was due to intense crab predation. In sheltered localities mussels grew more rapidly and therefore had a size refuge, whilst on the most wave-swept shores crabs were largely absent.

Apart from broad patterns of vertical zonation and local variability mediated mainly by wave action, each shore also exhibits small-scale patchiness. Barnacles and mussels settle preferentially in grooves and depressions where they presumably find local moisture and protection from physical abrasion. Waves and waterborne objects (e.g. stones, drifting logs) restrict some sessile forms to protected crevices. Biological interactions also create patchiness, often simply through the passive association between species. Mussel clumps, for example, provide a suitable habitat for numerous infaunal species whilst their shells afford secondary space for the attachment of epifauna. Macro-algae, such as the kelp *Hedophyllum*, provide a protective canopy and when this is removed, either experimentally or by natural disturbance, many obligate understorey algae which are adapted to lower light levels die back, whereas other rapidly-colonizing fugitive species (e.g. *Ulva, Porphyra*) prosper. On the Californian coast the limpets *Collisella digitalis* and *C. scabra* both occupy the high-shore but effectively partition their habitat on a small scale by their different behavioural responses. *C. scabra* lives on flat or gently sloping surfaces and by returning to a 'homescar' at low tide is effectively protected against desiccation. *C. digitalis* lacks a 'homescar' and is thus largely restricted to shaded, north-facing vertical slopes or crevices. Social behaviour such as gregariousness and territorial aggression exhibited by some limpets (e.g. *Lottia*) and clonal anemones (e.g. *Anthopleura*) also contribute to small-scale patchiness, as indeed does intensely localized predation and grazing.

Rocky-shore organisms are therefore limited in their distribution (and zonation) by a variety of environmental factors. Physical factors (e.g. desiccation, extreme temperature, salinity, turbidity) are generally limiting at the extremes of environmental gradients, whilst biological interactions become increasingly important wherever conditions are more benign. It must be emphasized, however, that factors rarely operate singly, and distribution patterns are almost invariably due to a complex interaction of many factors.

Disturbance, diversity and community organization

An important organizing principle of many space-limited communities is the existence of an approximate hierarchy in the competitive abilities of the primary space occupiers (see also p. 100). Such hierarchies, if left undisturbed, would lead eventually to extreme dominance by the competitively superior species (Dayton, 1971). However, moderate levels of physical disturbance (e.g. wave action, log impact) and *selective* predation (= biological disturbance) operating on species high in the hierarchy promote species diversity by interrupting the competitive process and creating patches of bare rock which can then be exploited temporarily by lower-ranking species. Low levels of disturbance may be insufficient to prevent competitive monopolization whilst high levels (e.g. ice scour, intense predation or herbivory) reduce diversity by acting adversely on all species (Figure 3.5). Predators which consume competitively dominant species are called *keystone species*. On many rocky shores a mussel is simultaneously the competitive dominant *and* the preferred prey for the principal predator (often a starfish) strongly suggesting that common mechanisms serve to organize rocky shore communities in different geographical regions.

The activities of the dominant species themselves may also contribute to the generation of bare patches of rock. Dense settlements of mussels sometimes lead to the formation of multilayered hummocks. As the underlying mussels are eventually smothered the hummock becomes increasingly unstable and is eventually torn away during storms. Similar intrinsic mechanisms have been described for other rocky-shore organisms (e.g. barnacles, sea-palms).

Succession is the orderly, predictable progression of a community

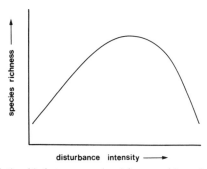

Figure 3.5   Relationship between species richness and intensity of disturbance.

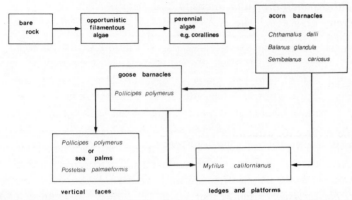

Figure 3.6   Succession of species in a Pacific coast intertidal mussel bed community.

towards a stable climax condition. A central requirement of the classical version of this concept is that early colonizing species (= pioneers) modify the environment so that it becomes suitable for species later in the successional sequence. Classical succession is not generally thought to occur in the rocky intertidal. In some cleared patches there may indeed be a definite sequence of species before the competitive dominant species reappear (Figure 3.6), but in other patches this does not occur. This is due in part to patch size and in part to the timing of clearance (disturbance) relative to the availability of larvae and spores. However, even where there is an apparently orderly sequence of species (especially in large patches) there is no evidence to suggest that pioneer species are *essential* for subsequent colonization by other species. Dominants such as mussels can colonize cleared patches *directly* providing their larvae are available. Pioneer species, however, are typically short-lived, competitively inferior, opportunistic species which, unlike most dominants, have early and extended reproduction. Availability of larvae is therefore virtually guaranteed and these exploit unpredictably freed patches of rock. The sequence of colonization is therefore not 'community-controlled' but due mainly to the different life-history characteristics of the colonizing species. Species which confer much of the structural integrity on the mature community tend to be long-lived species such as perennial algae, barnacles and mussels. Once these have become established they can modify the environment and provide access for other species (e.g. obligate understorey algae). Many of the later stages of algal succession are mechanically and chemically resistant to herbivorous grazing.

Rocky shores can perhaps therefore be visualized as mosaics of patches which, depending on the intensity and time of disturbance, will be at different stages in the successional sequence towards competitive dominance.

Productivity and trophic interrelationships

Rocky coastlines provide a unique habitat for plant growth, offering a stable surface for attachment (algae have holdfasts, not roots) in shallow, well-illuminated water which is often turbulent and rich in nutrients. Macro-algal communities, which in terms of biomass are generally dominated by kelps (e.g. *Laminaria, Ecklonia, Macrocystis*) and fucoids (e.g. *Fucus, Ascophyllum*) occur mainly in cooler water but extend into subtropical latitudes particularly in regions where there is upwelling of cold bottom water. Although total production is small compared with world phytoplankton production, on an area-for-area basis coastal algae are usually far more productive. Net primary productivity in kelp beds, for example, is often around $1000\,g\,C\,m^{-2}\,y^{-1}$, comparable therefore with other highly productive ecosystems such as tropical rain forests and coral reefs. The giant kelp *Macrocystis pyrifera* is probably one of the fastest-growing of all plants (up to 45 cm day^{-1}) and often reaches a total length of 60 m. Gas bladders situated at the base of the fronds keep these enormous kelps in the surface water.

Kelp fronds are essentially moving belts of tissue, and growth is measured by following holes punches between the stipe and frond adjacent to the meristematic tissue (Figure 3.7). As new tissue is added basally, old tissue erodes distally, so that the frond is gradually replaced; where growth is particularly rapid, total frond replacement may occur several times each year. This constant erosion generates large quantities of POM which is subsequently attacked by micro-organisms and rapidly enters the detrital food chain (Chapter 4). In addition to the production of new biomass, kelps also exude large amounts of DOM. After flocculation by bacteria this too enters the detrital food chain.

The major herbivores in kelp beds are sea-urchins. These feed mainly on detritus, detached drift weed or on newly established plants, but when released from predation they can be responsible for mass destruction of kelp. This occurred along parts of the Californian coast following the virtual extermination of the sea otter and in Nova Scotia as a result of lobster overfishing. Apart from urchins and a few gastropods there are few grazers of established kelps and most of their production eventually

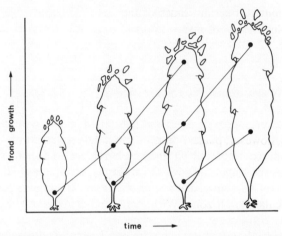

Figure 3.7   Kelp growth measured from holes punched into the base of the frond.

becomes detritus. This supports a diverse community of suspension feeders which includes sponges, mussels, holothurians and ascidians; these in turn are consumed by predators such as starfish, lobsters and fish. Uprooted plants cast on to the shore are degraded by myriads of amphipods and isopods. These complex trophic relationships are driven by energy subsidies in the form of wind and waves. Upwelling, generated by offshore winds, provides a supply of nutrient-rich water for the autotrophs. Strong wave action abrades the frond tips and breaks off fragments of weed or even whole plants, thereby creating space for new plants and providing a constant supply of detritus. Waves also restrict grazing by keeping the plants in constant motion.

Morphologically fucoids are more complex than kelps, and consequently their growth is more difficult to assess. Although highly productive ($c.$ $300\,g\,C\,m^{-2}\,y^{-1}$ net) fucoids are essentially intertidal, and exposure to various environmental stresses (e.g. desiccation, extremes of temperature and salinity) often results in much of this production being 'lost' as DOM. Very few herbivores feed on established plants, and most fucoid production, like that of kelp, eventually enters the detrital pathway, either directly as fragmented algal particles (POM) or indirectly as DOM. Herbivores can, however, have a dramatic effect on algal distribution on rocky coasts by removing spores or young plants; experimental removal of limpets, for example, usually results in luxuriant growths of algae even in areas where these were previously absent.

The major factors influencing macroalgal production are nutrient availability, the degree of water movement and light. Growth may be nitrogen-limited at certain times of the year. Consequently, in order to utilize the high levels of nutrient during autumn and winter (p. 26) some species (e.g. *Laminaria saccharina*) have become adapted to photosynthesize at low temperatures and therefore grow most vigorously during winter when light levels are least favourable and sea temperatures may be close to zero. Vigorous water movement over the frond surface constantly renews the supply of CO_2 and nutrients, facilitates their diffusion by reducing boundary layer resistance and removes metabolic wastes. In *very* turbulent conditions, however, plants become increasingly stressed and devote more of their resources to producing strengthening tissue. As light penetrates sea water, different wavelengths are absorbed differentially (Chapter 2). In general, therefore, seaweeds have photosynthetic pigments best suited to their particular depth zone. With increasing depth, however, algae can increase both the total concentration of pigment and the relative proportions of the component pigment types. Such *chromatic adaptation* effectively enables the photosynthetic mechanism to become saturated at lower light intensities and adapts the algae to changing spectral distributions of light. Intertidal algae acclimatize either as 'sun' or 'shade' forms by changing pigment concentration only. Light requirements may vary with developmental stage and this may be an important factor setting lower zonal boundaries.

Since most food webs are exceedingly complex it is often convenient to examine smaller units or *subwebs*. Two such subwebs for rocky shores along the Pacific coast of North America are illustrated in Figure 3.8. The northern subweb is strongly regulated by the starfish *Pisaster*. Removal of this keystone species reduces the diversity of primary space occupiers and sometimes results in a virtual mussel monoculture. Snails of the genus *Nucella* are trophically more specialized than *Pisaster*, feeding principally on barnacles and mussels. The southern subweb supports proportionately more predators. The top-ranking carnivores are the starfish *Heliaster* and the gastropod *Muricanthus*. Below these, the next two trophic levels are occupied by several carnivorous gastropods which prey mainly on herbivorous gastropods, bivalves and barnacles.

Other habitats

Algal epifaunas

The fronds of macroalgae which typify many sheltered rocky coasts provide a suitable and extensive surface for the attachment of numerous

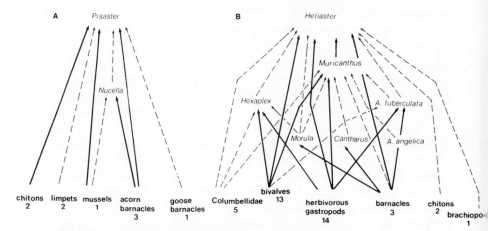

Figure 3.8 Trophic relationships in rocky shore subwebs on the west coast of North America. (*A*) Washington State; (*B*) Gulf of California. Major and minor pathways indicated by solid and dashed lines respectively: figures denote numbers of species. (After Paine, 1966.)

suspension-feeding invertebrates as well as shelter and food for many mobile species including fish. Two major factors are important for community organization: (1) larval selectivity in respect of acceptable environments, and (2) regeneration of space through plant growth (Seed and O'Connor, 1981). Prospecting larvae of many species can discriminate not only between different algal species, largely on the basis of their chemical attractiveness, but also between different regions within individual plants. Settlement is generally on to the younger, less heavily encrusted regions of the frond; this is clearly adaptive since it reduces competition and usually provides a substratum for the maximum period of time. How larvae identify the age gradients involved is uncertain but bacterial films and physiological gradients could provide the necessary cues. Plant growth renews the primary limiting resource (space) and enables many fugitive species to persist within these communities in which disturbance is often minimal. A further important feature of these communities lies in the nature of their competitive interactions where directional strengths vary spatially and seasonally, creating features of both hierarchies (p. 49) and networks (p. 100). Even without space regeneration, therefore, the stochastic (= chance) element of these interactions would probably preclude complete monopolization by potential competitive dominants. Many kelps have complex three-dimensional holdfasts which support numerous epifaunal and infaunal species.

Crevices

Crevices provide damp, protected microhabitats which frequently contain organisms not normally encountered elsewhere on the shore. Crevice faunas include many terrestrial air-breathing forms (e.g. insects, myriapods, mites) which utilize small pockets of air trapped within the crevice or beneath fine bristles covering their bodies. Factors influencing the zonation of organisms within crevices include humidity, light intensity, water circulation and the amount and nature of accumulated sediment (Figure 3.9) all of which in turn depend largely on shore level and size of crevice. At low tide, crevices provide refuge for many transient species such as crabs and isopods (e.g. *Ligia*). Environmental gradients, in particular in light and water circulation, are also important in determining the zonation of organisms in sea caves.

Tidal rapids

Areas which are protected from wave action but which experience the benefits of clean, fast-flowing water are especially species-rich. Such tidal rapids occur in channels between adjacent land masses especially where the entrance to a bay or sea-lough is narrow by comparison with the larger area beyond. Their floras are usually dominated by luxuriant growths of large algae (e.g. *Himanthalia, Halidrys, Laminaria*), their faunas by attached suspension feeders.

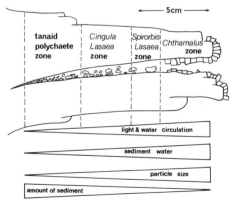

Figure 3.9 Profile of a generalized crevice illustrating the distribution of the dominant organisms and gradients in the physical environment. (After Lewis, 1964.)

Boulders

The upper well-illuminated surfaces of intertidal boulders characteristically support communities similar in composition to those on adjacent bedrock. The under-boulder community, by contrast, which is protected from extreme desiccation and wave action, is dominated by shade-loving (= *sciaphilic*) species in which sponges, bryozoans, ascidians, hydroids and tubeworms are particularly prominent, the precise species composition varying according to the degree of tidal exposure. Specialist predators, mainly gastropods and echinoderms, are also frequently present; so too are crabs, many of which (e.g. *Porcellana platycheles*) have flattened bodies for life in restricted spaces. Physical disturbance appears to be particularly important in organizing boulder communities. Small boulders (and cobbles), especially in areas of strong water movement, are regularly rolled over and consequently only the most opportunistic species can become established. Large stable boulders resist wave overturn, thus enabling a few competitively superior species to dominate the community. The most diverse faunas occur on boulders experiencing moderate levels of disturbance, and in this respect boulder communities resemble other types of rocky-shore communities (p. 49). Wherever boulders are embedded in anoxic sediment (e.g. in extreme shelter) colonization of the undersurfaces effectively ceases.

Subtidal rock

Subtidal communities, like intertidal communities, are vertically zoned but here the zones are usually broader and less pronounced. Moreover, the full sequence of organisms is often truncated wherever rock gives way to expanses of sediment. Schemes for classifying these communities have been proposed (e.g. Hiscock and Mitchell, 1980) but these generally lack the wide-scale applicability of those described for rocky shores (p. 40). Desiccation is clearly unimportant here and zonation is determined by light, water movement and biotic factors. Whereas the intertidal fauna consists mainly of solitary forms (e.g. barnacles, mussels) most organisms occupying primary space subtidally are colonial (e.g. sponges, ascidians, corals). Dense aggregations of solitary forms (e.g. oysters, vermetids, tubeworms) do, however, occasionally form extensive subtidal reefs. Colonial organisms occur in various growth forms but convergent evolution has led to the recurrence of a few morphologies (e.g. sheets, runners, mounds) amongst taxonomically unrelated groups. These do reproduce sexually to meet their needs for dispersal stages (larvae) and to maintain genetic variability, but it is their asexual proliferation ability that largely underlies their great success subtidally. The adaptive significance of solitary and colonial strategies is considered by Jackson (1977).

In cooler waters dense growths of algae (especially kelps and erect and crustose red algae) occupy the more gently shelving bottoms wherever there is adequate light; on steeper rock walls and at greater depths these are replaced by encrusting animals. Tropical communities are dominated by animals, particularly corals, though crustose algae may be abundant at shallower depths (Chapter 6). In polar regions the shallow subtidal is largely devoid of permanent communities due to the scouring effects of ice, though transient species may appear in areas which become temporarily ice-free. At somewhat greater depths there may be a diverse community dominated by sponges and finely regulated by predatory starfish. Thus, patterns and processes of community organization in the rocky subtidal are similar to those operating in the intertidal zone. For further details of selected aspects of rocky coast ecology, and an extensive bibliography, see Moore and Seed (1985).

CHAPTER FOUR

LIFE IN SEDIMENTS

The physical background

Sediments consist of particles which tend to become round due to abrasion, although initially most of them are angular. When the particles are packed together spaces between them occupy 30–40% of the sediment volume. If the particles were spherical the solid to void ratio would be 74:26. Thus there is a system of *interstices* which may be variously filled with water, air, detritus and organisms. The water in sediments may be derived or maintained by various means. On beaches the surface tension or capillary forces may retain water which has been derived from the groundwater table or from tidal input; there may be superficial or subterranean streams of fresh water as well as rain and evaporation, all of which will modify salinity. Some fresh water may emerge through the sea-floor at considerable distance offshore. However, when present, fresh water tends to flow over the surface of beach sediments which are thus protected from extreme salinity fluctuations. Sediments also provide some insulation from temperature change (Figure 4.1) providing that the organisms can avoid the immediate surface. The specific heat of damp sand is usually 0.1–0.3 of that of water. The granulometric composition of sand and mud affects various other parameters, including the total pore volume, as well as pore dimensions. An admixture of fine particles such as silt or sponge spicules alters the pore space radically and this in turn alters the sediment's drainage or percolation characteristics.

The oxygenation of sediments falls rapidly when there is more than 30% of material of less than 200 μm diameter. Intertidal organisms may experience severe oxygenation problems, for example during emersion they are often dependent on the interstitial or burrow water for both respiration and waste disposal. Nearly all sediments become anaerobic at some depth from their surface determined by the local hydrodynamic regime, the input of organic matter and the sediment biota. In muds this depth is often very

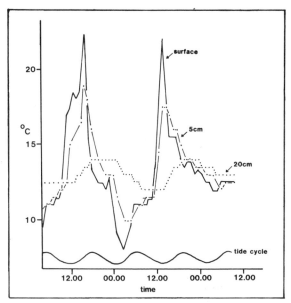

Figure 4.1   Temperature attenuation with depth on a sandy beach. Data from Kilroot, N. Ireland, August 1979. Subsurface changes are smaller and later than at the surface. (After Wilson, 1983).

close to the sediment surface, in exposed-beach sands it may be a metre or more deep. Water circulation in tidal and subtidal sands can be considerable. On exposed beaches it has been estimated that 20–200 metric tonnes of water pass through each metre length of beach per day due to the swash and backwash drainage percolation (Riedl, 1971). The mean drainage speed through the interstices during these processes reaches $300\,\mu\mathrm{m\,s}^{-1}$ and any particle in the water has three chances of striking a meiofaunal animal (p. 74) during passage. Water is also pumped in and out of subtidal sediment ripples due to pressure change induced by waves.

The 'structural property' of sand and mud changes radically with water content, particularly in sediments of the 180–250 μm median particle-size range. In coarse sediments water tends to move away from any site of mechanical disturbance, the sediment becomes more compact and difficult to penetrate. Beach sands which show hard whitish footprints are of this nature. This property is known as *dilatancy*. In finer sediments water tends to move towards disturbance and the sediment fluidizes; it is therefore easy to enter but hardens on removal of the disturbance. It is thus particularly good for burrowers and effectively removes footwear. Such sediments are called *thixotropic*.

Prokaryotes and interstitial chemistry

The number of eubacteria in sediments increases as the particle diameter decreases. In part this is due to the increased particle surface area since over 90% of sediment bacteria are attached to sediment and detritus particles. Unfortunately there are still very few comparable counts of bacteria because of the wide variety of determination methods. Taking the top 5 cm depth, bacterial numbers of $200 \times 10^6\,\mathrm{g}^{-1}$ of sand and $5\,000 \times 10^6\,\mathrm{g}^{-1}$ of mud, or bacterial weights of $5.5\,\mathrm{g\,m}^{-2}$ and $57.0\,\mathrm{g\,m}^{-2}$ respectively (and some fourfold greater values) have been reported. Bacterial productivity is probably about $100\,\mathrm{g\,C\,y}^{-1}$ per gram bacterial biomass (i.e. $P\!:\!B$ ratio $= 100$). Cyanophytes are also present, sometimes in considerable quantity and at some depth in the sediment, where they may be existing heterotrophically. Some species such as *Beggiatoa* (Figure 4.2) and *Oscillatoria* produce surface mats and are important in nitrogen fixation,

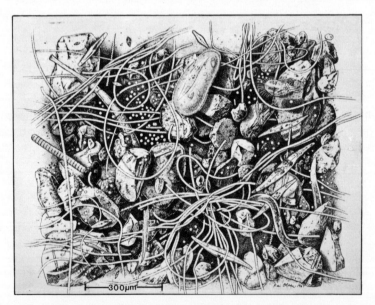

Figure 4.2 Sulphide-rich sediment surface with filaments of *Beggiatoa* and rounded *Thiovolum* bacteria together with ciliates, diatoms and a nematode. (From Fenchel, 1969.)

for example in tambaks (Chapter 9). In mangals (or mangrove communities: see Chapter 7) their planktonic hormogonia are of importance in food chains.

Hydrodynamic regimes which allow sediment settlement are also likely to allow deposition of organic particles. These may well have undergone some bacterial decomposition beforehand but will be colonized by further bacteria on reaching the sediment. Bacterial respiration, like any other, requires an oxidative source. In the sediment surface this can be dissolved oxygen. However once this is utilized any other organic matter can only be decomposed using other oxygen sources such as nitrate, sulphate and carbon dioxide or bicarbonate ions (Figure 4.3). These are reduced progressively to ammonia, hydrogen sulphide and methane. Production of the latter two compounds involves fermentative processes in which complex organic molecules are broken down into simpler constituents, thus anoxic interstitial water is rich in amino acids and carbohydrates. It usually also contains substantial amounts of phosphate and silicate. Because of the imbalance of ions involved in sedimentary reduction and oxidation processes, sediments have an electrical charge. This is known as the *redox potential* or Eh. The sediment Eh profile usually has a characteristic shape (Figure 4.3) with a plateau in the region where there is a shift from positive to negative charge. This is called the redox-potential discontinuity (RPD) layer and more or less corresponds to the change from oxidizing to reducing conditions (Fenchel, 1969).

Sediment sulphur cycling

Sulphate is the third most common ion in sea water (following sodium and chloride) so it is not surprising that sulphate reduction plays a dominant

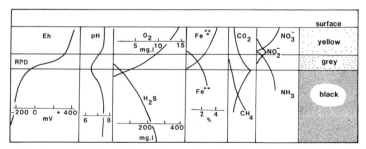

Figure 4.3 Some physical and chemical properties of sandy sediment. Redox potential is a general guide to oxygenation state and shows a plateau, the redox potential discontinuity (RPD) layer, as conditions become anoxic. (After Fenchel and Riedl, 1970.)

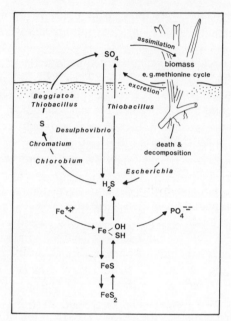

Figure 4.4   The sulphur cycle of marine sediments showing some of the main prokaryotes involved.

role in sediment chemistry. The bacterial/sediment sulphur cycle is outlined in Figure 4.4. Sulphates are used by plants which reduce them to sulphydryl compounds for their methionine cycle. Sometimes prokaryotes involved in the oxidation of sulphur and sulphides occur in sufficient quantity to colour sediments; *Chlorobium* and *Chromatium* belong respectively to the green and purple sulphur bacteria. The latter is often visible as pinkish-purple patches in damp areas next to decaying seaweed. Both these forms use light energy to photosynthesize compounds from carbon dioxide and hydrogen sulphide. *Chromatium* is sensitive to damage by sulphydryl ions, whereas these are necessary for *Chlorobium*. *Beggiatoa* is a sliding filamentous cyanophyte and, like other forms such as *Thiothrix* and *Thiovolum*, tends to form films at a particular sulphydryl tension. They are often visible as white patches in areas of decaying weed or over dead animals left in aquaria; the colour is due to internal depositions of sulphur. Many of the sulphur-reducing forms are also important in nitrogen fixation. Most *Thiobacillus* species require oxygen to oxidize sulphur, sulphide or thiosulphate to sulphate. This generates energy which is used in the synthesis of organic compounds from carbon dioxide. Nearly all thiobacilli therefore require carbon dioxide, free oxygen and low-oxidation-state sulphur compounds; most of the latter are produced by anaerobic bacteria. Therefore the thiobacilli can only occur at the interface of anaerobic and aerobic conditions. *T. denitrificans* is unusual since it can exist anaerobically using denitrification energy to oxidize sulphur. *Desulphovibrio* is an obligate anaerobe existing at less than $+ 100 \, mV$. It is involved in the reduction of sulphur and of sulphate but has to 'rely' on other reducing organisms to provide a suitably low Eh milieu. Various iron–sulphur compounds are formed in anaerobic conditions and there is a shift from trivalent to bivalent (ferrous to ferric) iron. Their formation is accompanied by the reduction of ferric phosphate, and the phosphate thus solubilized may be released from the sediment by diffusion, turbulence or biological activity. Sulphide is a strong complexing ligand and, since many of the metal sulphides are insoluble, metallic ions tend to be 'locked' into anaerobic sediments.

Sediment Protista

Protistans make two fundamental contributions to sediment ecology, namely (1) through primary productivity which provides an autochthonous food source in intertidal and shallow areas, and (2) through heterotrophy which plays a major role in incorporating allochthonous material (previously metabolized by prokaryotes or fungi) into the benthic food web. Usually the majority of sediment primary production is due to pennate diatoms, such as *Nitzschia* and *Hantzschia*, and to dinoflagellates. Many of these species are motile and undergo periodic vertical migrations in the sediment which may lead to the surface changing colour to an olive-brown or greenish hue. In some species and localities there is an endogenous diurnal vertical migration controlled by day-length, but an additional tidal rhythm is superimposed by reaction to mechanical disturbance and wetting. Presumably the function is to maintain the diatoms in well-illuminated conditions except when it is necessary to avoid disturbance or desiccation. Some species counter the latter problem by secreting a mucilaginous sheath. Many benthic diatoms are photosynthetically efficient at low light intensities. Surface mud contains up to 5×10^5 diatoms cm^{-2}. Euglenoids (especially in estuaries) and chrysophyceans can also contribute significantly to primary productivity. Production values on boreal estuarine mud and on sand-flats are generally between $100–200 \, g \, C \, m^{-2} \, y^{-1}$. Exposed sand beach primary production is about $5–10 \, g \, C \, m^{-2} \, y^{-1}$.

Large numbers of heterotrophic flagellates and ciliates occur in muds and sands (see Fenchel, 1978) although the former, which are of increased importance in sediments of $< 100\,\mu\text{m}$ diameter, have been investigated in very few places. Ciliates may number up to $40 \times 10^6\,\text{m}^{-2}$ in fine sands of $150–250\,\mu\text{m}$ median diameter where they represent up to a fifth of the faunal biomass. Attenuated ciliates such as *Trachelocerca* and *Remanella* species are common in fine sands and are therefore sometimes termed microporal; their length often exceeds that of some meiofaunal species. Coarser sediments tend to have shorter wider 'mesoporal' ciliates. The majority of sediment ciliates feed on bacteria, flagellates and diatoms although some scavenge dead and moribund animals. There is usually a marked vertical zonation which is related to oxygenation and sulphide tolerance as well as to food distribution. Ciliates are fairly numerous in the surface floc of subtidal sediment and in algal and prokaryote mats. Some species living in anoxic layers carry a bacterial coat. Total ciliate metabolism in sediments may equal or exceed that of the macrofauna; furthermore, grazing by ciliates stimulates bacterial growth. Ciliates are predated by some turbellarians and gastrotrichs and probably by nematodes, as well as being ingested by macrobenthic deposit feeders.

Macrophytes

Few large algae occur on sediments. Where there is some admixture of stones or shell this may be used for attachment, for example by the bootlace weed, *Chorda filum*. There are some species, such as *Gymnogongrus linearis* of the North American Pacific coast, which are characteristic of rock which is intermittently covered by sand. In lagoons and other extremely sheltered areas some algae produce unattached morphs which cover the sediment at low water but float as the tide rises; these ecads include *Ascophyllum nodosum mackaii* and *Ectocarpus* '*distortus*' on European shores. There are also mat-forming filamentous species such as the widespread *Rhizoclonium implexum*.

Sea-grasses

Except in polar regions, photic-zone sand and mud communities are often characterized by sea-grasses (also termed eel- or turtle-grass). They are monocotyledonous angiosperms adapted for marine life both through their physiology and morphology (see McRoy and Helfferich, 1977). The most obvious characters are the extensive rhizome and rooting system and, generally, the very flexible strap-like leaves. There are 49 species, 11 belonging to the genus *Zostera*. The latter are mainly boreal and temperate species and, since they have little capacity for upward rhizome growth, are limited to areas where the deposition and erosion of sediment is more or less in equilibrium. However, presence of the sea-grass slows water movement across the sediment, which leads to more deposition of finer particles than in the surrounding areas. *Zostera* roots penetrate the anoxic layer of sediments but sometimes exhibit a surrounding yellow-brown halo which may indicate local oxygenation. The root hairs attach to hydrotroilite particles. Phosphate and ammonia are released from the sediment via the *Zostera* plants. Most Atlantic populations of *Z. marina*, the common low- to sub-tidal species of northern temperate areas, were subject to a 'wasting disease' in the 1930s. This was thought to be due to fungal infection but it has since been established that even healthy plants contain such fungi; it is probably that the widespread death of the grass resulted from a series of unusually warm summers.

Tropical sea-grass beds on mud, sand or coral rubble often contain members of several different genera such as *Thalassia* (turtle-grass), *Syringodium* and *Cymodocea*. In some warmer regions there are subtidal beds of more rigid sea-grasses, such as *Amphibolion* in south-western Australia. Warm temperate waters of the Mediterranean and of Australia support subtidal *Posidonia* meadows. The rhizomes of *Thalassia* have some capacity for vertical growth but this is much greater in *P. oceanica* and *P. australis*. *Posidonia* rhizome systems can grow upwards at a rate of a few millimetres each year to produce a raised platform or *matte*. Some of these are 10 m thick and thus must be thousands of years old. This accretion process can be the climax stage of a succession leading from rock, through encrusting epilithic algae and forms such as *Padina*, to a sandy subtidal meadow overlying the original rock.

The measurement of primary productivity of sea-grasses presents various problems and it is often not clear from the literature whether or not values include the contribution from epiphytic diatoms. However, many of the published figures are in the range of $0.5–5.0$ g C m^{-2} day^{-1}. Values of three to four times this (matching terrestrial cereal production) have been reported. The mean annual values reach 1 kg C m^{-2} in the tropics and 0.5 kg in temperate areas. The leaf area of sea-grasses provides considerable space for attachment of bacteria, diatoms and other forms. The surface area provided by *Zostera* and *Thalassia* can reach 20 m^2 per m^2 of sediment surface. There is a rapid loss of dissolved organic matter from sea-grass material after it has abraded or broken off, but decomposition then occurs rather slowly with a weight loss of about 10% per week. Each gram (dry weight) of sea-grass detritus can contain $50-500 \times 10^6$ heterotrophic protists which between them consume $200-2000 \times 10^6$ bacteria per hour.

Sea-grass communities

Kikuchi and Pears (in McRoy and Helfferich, 1977) have reviewed the fauna and flora of sea-grass meadows. They recognize four main sub-habitats. These are (1) the leaf epiphyton comprising of microflora with associated small animals including nematodes, polychaetes and crustaceans, together with sessile fauna, such as hydroids, sea-mats and anemones, and vagile forms including snails, echinoderms and small fish; (2) stem and rhizome biota which include various polychaetes, amphipods and bivalves; (3) species swimming among the leaves including fish, especially wrasse, cephalopods such as *Sepiola* and crustaceans; and (4) sediment fauna, although this may differ little from that of the surrounding benthos.

Bottom communities

The term 'community' does not have an agreed biological definition although it is in regular use by ecologists. Arguments about its use stem from the separate histories of terrestrial and marine biology. Many terrestrial ecologists have used the term *sensu lato* to describe a group of organisms from a particular habitat, but others have evolved a strict definition which requires a community not only to be distinct from neighbouring communities but also, given radiant energy, to be self-sustaining. The great Danish marine biologist Petersen, however, used the term to describe groups of species which occurred together regularly in bottom grab samples and which were usually present in large numbers. Petersen's concept was extended by various marine workers, especially Thorson (1957), who had noted the occurrence of functionally similar (and often congeneric) species in soft sediments of similar composition around the world (Figure 4.5). He termed the biota of these areas 'parallel bottom communities' and named them after their dominant species. The functional concept cannot be applied to the *sensu lato* communities, and the *sensu stricto* self-sustaining concept cannot be applied in most marine environments. Various workers have stressed the physical background, parti-

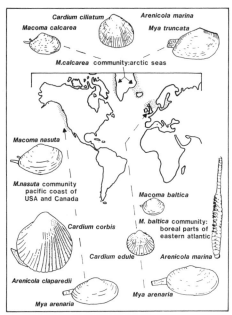

Figure 4.5 Parallel bottom communities in shallow muddy sand. (After Meadows and Campbell, from Thorson, 1957).

cularly depth and substratum, in naming communities, using the dominant genera or species to name *facies* or constant biota to name *associations*. The term 'biocoenosis' is also used for the 'assemblage' of species in a particular habitat. The resulting arguments have many of the characteristics of schismatism, and many faunistic investigations have been aimed merely at listing and naming facies rather than trying to understand causation. However, the term 'community' remains a pragmatic description for the broadly similar biota from broadly similar habitats. This is essentially the same use as in the environmental classification system (Table 4.1) proposed by Erwin (in Earll and Erwin, 1981).

However, to complete this hierarchy a super-taxon of *biogeographical provinces* is required (Figure 4.6). The most important factor controlling this scale of distribution is water temperature. The major geographical boundaries occur at intervals of about 5–15°C or a corresponding scale of 10–30° of latitude. The provinces are best defined for intertidal fauna and flora but correspond fairly well to the geographical distribution of oceanic plankton. They also hold for subtidal and other shelf benthos (see Ekman,

Table 4.1 A system for classifying inshore assemblages (from Erwin, in Earll and Erwin, 1981)

Taxon	Area of decision	Hierarchy
1. Salinity regime	Prevailing salinity hypersaline; euhaline; mixohaline	Different depth regimes may represent the same *salinity regime*.
2. Depth regime	Depth zone, e.g. upper infralittoral or depth band, e.g. 0–5 m below datum	Different situations may represent the same *depth regime*. Other depth regimes exist, each with their own range of variable situations.
3. Situation	Stability regime, e.g. in the sublittoral fringe—very exposed, exposed, semiexposed, sheltered, very sheltered	Different formation types may represent the same *situation*. Other situations exist, each with their own range of variable formation types.
4. Formation type	Gross nature of substratum, e.g. bedrock	Different communities may represent the same *formation type*. Other formation types exist, each with their own range of variable communities.
5. Community	Habitat(s) available, e.g. on bedrock-vertical face, horizontal face, crevice, etc.	Different facies may represent the same *community*. Other communities may exist, each with their own range of variable facies.
6. Facies	Major constant obvious biotic elements	Different associations may represent the same *facies*. Other facies may exist, each with their own range of variable associations.
7. Association	Constant biological assemblages	Within any *association* there is variation of populations and individuals.

1953; Briggs, 1974). The occurrence of communities within these provinces will obviously be controlled by several factors including salinity (Chapter 5), depth and substratum. Depth has an obvious effect on the composition of rock communities because of the photic requirements of algae, but the light regime has much less influence in sedimentary environments. In these, temperature regime and sediment stability are of prime importance. Each of the major sediment grades of gravel, sand and mud will have at least one community with three or four fairly recognizable facies. Thus in European shelf waters there is a gravel community with typical bivalve-named facies, i.e. (1) inshore, *Dosinia exoleata* or *Venus verrucosa* if it is muddier; (2) further offshore, *V. fasciata*; and (3) in deeper water, *Venus casina*. The lancelet *Branchiostoma* (*Amphioxus*) often occurs in this community.

Much of the original impetus for community studies came from fisheries investigations. Thus, although studies after World War I proceeded more by grab than by trawled gear, it was always envisaged that communities

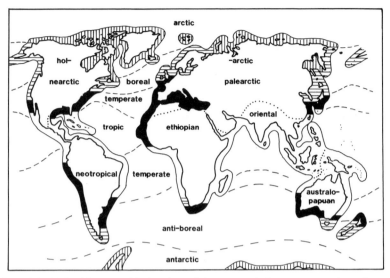

Figure 4.6 Major biogeographical realms of the oceans, inshore waters and continents. Littoral provinces shaded as follows: vertical hatching, Arctic and Antarctic; horizontal hatching, boreal and anti-boreal; solid shading, warm temperate and unshaded, tropic.

contained both an *infauna* (sometimes dignified with Greek habitat names such as 'endopsammon' or 'endopelos' for that of sand and mud respectively) and an *epifauna* ('epipsammon', 'epipelos' etc.) including fish and other forms living on the sediment surface. Complications arose since some species, e.g. *Nephrops*, live in both manners and because the early investigators usually sieved their grabbed sediment through 1- or 2-mm meshed screens, thus losing the smaller animals. Later investigators showed that animals which passed through such sieves represented a significant portion of the benthos. They were termed *meiofauna* and the idea of size categories, the macro-, and meio- and microfauna (protozoans) was established. Since the meiofauna of sand was the first to be extensively investigated but did not fit into the approved endopsammal habit of displacing sediment by burrowing or tube construction, a third special habitat category, the *mesopsammon*, was designated.

Some macrofaunal representatives

Members of the intertidal to shallow-water mud-flat community (Figure 4.5), with some additions, will serve to illustrate several aspects of infaunal biology. The Polychaeta, Mollusca and Crustacea are usually well represented in sedimentary communities, as are the Echinodermata, although they are not so common intertidally.

Polychaeta

Partly as the outcome of World War II (the evacuation of part of the University of London to Bangor, North Wales) the lugworm *Arenicola marina* became one of the most-studied marine animals. The worm in its burrow (Figure 4.7) makes periodic excursions to defaecate at the tail shaft entrance. It returns to its feeding position where it irrigates the burrow by peristaltic contractions of the body wall passing from tail to head. Laboratory studies have shown that the defaecation/irrigation rhythm is controlled endogenously, and corresponding rhythmic activity is even undergone by small strips of body wall providing that there is some attached nerve cord. A separate oesophageal rhythm controls feeding activity (Wells, 1966). The worm ingests sand but there is argument about its main source of organic matter. Recent evidence suggests that bacteria originally brought into the head shaft by sand movement are the main source. This might represent a *gardening* phenomenon (see Hylleberg, 1975) in that bacterial growth on sand in the head shaft could be encouraged by irrigation and exudates. *Arenicola marina* haemoglobin

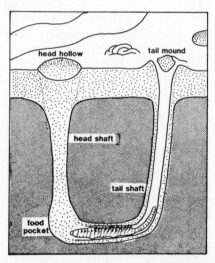

Figure 4.7 The lugworm *Arenicola* in its burrow. Tail shaft width exaggerated and length lessened; heavy stippling represents RPD and anaerobic sediment. (After Wells, 1966.)

increases the oxygen capacity of the blood tenfold and saturates at very low partial pressures. The worm also has a substantial capacity for anaerobic metabolism, possesses a glycogen energy store and has a sulphide detoxification system.

Terebellid polychaetes are also a common component of muddy shores. They are microphagous, using their numerous unbranched tentacles to gather small food particles from the sediment surface (*deposit feeding*). However, at least one species, (*Lanice conchilega*) can switch to *suspension feeding* by extending its tentacles into water currents carrying plankton. Some normally predatory worms such as *Nereis diversicolor* can suspension-feed by producing mucus nets. However this method is more fully developed in the chaetopterids. Some sedimentary polychaetes form small reefs by building tubes from particles taken from the sea-floor or caught in suspension. The settlement behaviour of *Sabellaria* is related to this mode of life since most of its larvae will settle only if they come into contact with the binding material used for tube construction by already settled young or adult worms.

Mollusca

The bivalves are essentially filter feeders—sedimentary forms possess siphons which enable either suspension feeding (as in *Cerastoderma* (= *Cardium*) and *Mya*), or deposit feeding (as in *Macoma* or the tellins which occur in sandy sediments). Young *Cerastoderma edule* are eaten in prodigious quantities by fish and by shore-birds. Siphons in this near-surface-living form are short and the shell is fairly thick. In *Mya*, which lives at some depth in mud, the siphons are long, fused and extensible and the animal can 'afford' a more fragile shell. In general, mud community animals tend to be restricted to the near-surface layers and have a thinner shell or tegument than their sand counterparts. Several bivalves have a substantial capacity for anaerobic respiration. This helps them to survive temporary burial and, on beaches and in estuaries, may help the animals survive freshwater influx, as is the case in the oyster *Crassostrea virginica*. In a few species, such as the Icelandic clam *Cyprina* (*Arctica*), anaerobiosis is self-induced by frequent periods of burrowing; the reasons for this are unknown. In several parts of the world exposed beaches are inhabited by surf clams (*Donax* spp.). These react to vibration on flood tide by emerging from the sand, and are thus carried up the beach by wave swash. They then reburrow at the new shore level. A reverse migration occurs in response to drainage currents during ebb tide and the clams are carried back down the beach. In this manner the clams maintain their position in the region of maximum food input.

Several prosobranch molluscs are infaunal or epifaunal. These include various predatory forms such as the Naticidae or necklace shells, so called because of their choker-like egg masses. They prey on bivalves which are killed after having a neat round hole bored through the shell. The tropical helmet shells prey on sea urchins, sometimes via the latters' anus; the Indo-Pacific *Cypraecassis* is used extensively in production of cameos. The commonest snail in Europe is probably *Hydrobia ulvae,* a small (< 5 mm tall) species which occurs in enormous densities on mud and muddy sand. It is extensively preyed on by ducks and other birds (Chapter 10), and hosts various helminth parasites such as *Cryptocotyle jejunum* which completes its life cycle by passing through gobies and redshank. The eggs, which are passed out in redshank droppings, only hatch if eaten by *Hydrobia. H. ulvae* often makes a habit of coprophagy, since it feeds extensively by burrowing through sediments which are partly composed of its own faeces. These contain significant quantities of bacterial protein rich in nitrogen which has been fixed by the bacteria as they oxidize the high carbon content left in the faeces after passage through *Hydrobia.* This is one of Nature's best attempts at a perpetual motion system. The mud-flat snail *Nassarius obsoletus* has been the subject of various studies on the United States' eastern seaboard. The larvae have a settlement response to a water-soluble substance from the sediment and the species is gregarious. It apparently migrates towards decaying-animal food sources but is generally regarded as a detrital or microalgal grazer. In some areas its abundance alternates with that of the tube-building amphipod *Ampelisca.* In spring and summer the tubes are so abundant that the snail's grazing is restricted, but autumn storms lead to the tubes being stripped from the sediment, thus freeing it for recolonization by *Nassarius.*

Crustacea

Amphipods, isopods and decapods are often common in or on soft sediments. In intertidal mud and sandy mud the amphipod *Corophium* or related forms occur in considerable quantity. The biology of two species, *C. volutator* and *C. arenarium,* has been studied in Great Britain. *C. arenarium* occurs in somewhat sandy areas, preferring a particle size of 44–104 μm. This, and its preference for aerobic sand, have been demonstrated in laboratory experiments in which *C. volutator* prefers burrowing into sediments of < 44 μm especially if they have been anaerobically conditioned. *Corophium volutator* is a deposit feeder and constructs small burrows into which it rakes surface detritus with its long antennae. In at least some areas there is a tidal migratory response similar to that of surf clams. The amphipods of sand beaches often exhibit vertical zonation down the shore.

In warm temperate and tropical areas, crabs are often the most conspicuous components of soft-shore fauna. Some species, such as *Emerita talpoides*, filter food material from water via their long feathery antennae. Some decapods, such as *Callianassa*, are modified for burrowing life by their general elongation and by modification of the swimmerets to produce circulation of water.

Echinodermata

The irregular or heart-urchins such as the sea-potato *Echinocardium* and the sand-dollar *Mellita* have spatulate ventral spines and specialized tube feet for digging; most appear to be deposit feeders. Sea-cucumbers are often members of sediment communities. Some species feed by ingesting sediment, either selectively (i.e. choosing particular particles) or non-selectively, and other species are filter feeders. Many sea-cucumbers can spew their guts in response to predatory attack, but they have rapid regeneration powers. There are numerous sediment starfish and brittle-stars. Several of the latter, as well as heart-urchins, have been used in naming facies. Brittle-stars often occur in dense concentrations on subtidal sands. Sediment echinoderms often have pointed rather than suckered tube feet, the latter being of little use for attachment to sediment.

Bioturbation

Deposit-feeding macrofauna can have profound effects on sediment stability, particularly in mud where they are relatively more common than in sand. The sediment turnover rates may be considerable—for example it has been calculated that the dense populations of polychaetes which occur in some low-tide terraces can process up to $2500 \, \mathrm{kg \, m^{-2}}$ of beach in a year. The enormous enteropneust *Balanoglossus gigas* processes about $90 \, \mathrm{kg \, m^{-2} \, y^{-1}}$ in some Brazilian beaches. Tropical holothurians can process similar amounts per individual. Some deposit feeders which ingest below, but defaecate at, the surface (see Figure 4.8) have been termed *conveyor-belt species* and have a particularly important effect in returning organic matter and nutrients from anaerobic to aerobic conditions where

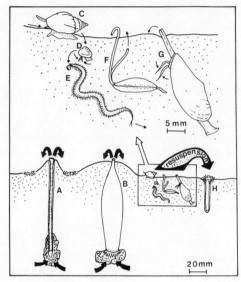

Figure 4.8 Methods of mixing and recycling of sediment by deposit-feeders: (*A*) maldanid polychaete; (*B*) holothurian; (*C*) gastropod (*Nassarius*); (*D*) nuculid bivalve (*Nucula* sp.); (*E*) errant polychaete; (*F*) tellinid bivalve (*Macoma* sp.); (*G*) nuculid bivalve (*Yoldia* sp.); (*H*) anemone (*Cerianthus* sp.). Oxidized mud lightly stippled, reduced mud unstippled. (From Rhoads, 1974.)

they are more easily available to other benthic species (Rhoads, 1974). It has also been shown that secretions, for example from some polychaetes, may help stabilize fine sediments.

Meiofauna

Sediment meiofauna is of interest for a number of reasons. These include (1) the illustration of relationships between morphology and habitat, (2) the linking of micro-floral and -faunal processes to the macrofauna, (3) its spatial distribution and (4) its possible evolutionary significance.

Sand meiofauna

Some species of the sand meiofauna are illustrated in Figure 4.9. The most numerous group is the nematodes, followed by copepods and/or gastrotrichs and turbellarians. Nearly all other invertebrate phyla have at least some interstitial species. In sand, nematode density can reach

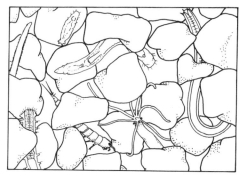

Figure 4.9 Some sand meiofauna, grain diameter about 0.5 mm. Shown from bottom left are a gastrotrich, a copepod, the cnidarian *Halammohydra*, a nematode, a seriate turbellarian, the smallest known annelid *Diurodrilus* and a ciliate.

$1\text{--}3 \times 10^6 \, \mathrm{m}^{-2}$; since numbers may be four or more times greater in mud and since nematodes are also of some importance in phytal systems, Platt and Warwick (in Price *et al.*, 1980) have stated that energetically nematodes may be 'the most important single metazoan group in many shores' ecosystems'. Although nematodes show various morphological adaptations, their long thin shape is already well suited to interstitial life. Some idea of food preference can be gained from their mouth structure; in sand particle-epigrowth feeders and predatory or omnivorous forms predominate. Some common sand meiofaunal adaptations are the following:

(1) Either attenuation, for progressing through interstices, or leaf-like form for moving around individual sand grains.
(2) Possession of tactile sensoria, photoreceptors and statocysts. These are concerned respectively (*a*) with sensing interstitial dimensions—in experiments many species can distinguish between various grades of sand and enter that closest to their natural habitat; (*b*) with photonegative response to keep out of the immediate surface layer of sand in which there are physical and biological dangers; and (*c*) with orientation, because the majority of environmental gradients in sand are vertical.
(3) Extensive ciliation or body flexion for locomotion.
(4) Appendage reduction—in several coarse sediment copepods the limbs are so flexible (to allow passage through the interstices) that they cannot be used for locomotion, and in the nudibranch mollusc *Pseudovermis* the cerata (papillae holding stinging cells derived from *Halammohydra*) do not protrude.

(5) Possession of adhesive tubules or similar organs—in several phyla these are duo-glandular with only two cells, one producing an adhesive and the other a releaser.

(6) Integumental cuticular structures or internal spicules—these are common in meiofauna from mobile substrata such as subtidal sand waves, and are thought to protect against mechanical damage.

(7) Hermaphroditism, suppression or reduction of larval stages and provision of brood care—these features compensate for the inability to produce large numbers of gametes which arises because of the lack of 'spare' metabolic capacity or body space.

Meiofaunal organisms are very patchily distributed within the sediment. The pattern of occurrence often follows the negative binomial or Neyman A (clumps of clumps) distribution. The reasons for this are not fully understood. Patchiness of food, in particular the predilection for certain bacteria, and a gregarious response may be involved. The latter is attributed to the difficulty of ensuring reproductive encounters within the three-dimensional maze of sand interstices. Gregariousness and bacterial attractiveness have been demonstrated by various laboratory experiments (see Gray, 1981). This and other evidence suggests that sand meiofaunal communities may be partly structured by chemical responses to ectocrines and other organic compounds. There is usually a clear vertical zonation of species within sand, and two or more congeneric species often occur within single core samples. Some species, families and even one phylum, the Gnathostomulida, appear restricted to anaerobic habitats in or below the RPD layer; it has been suggested that this habitat and its biota (together termed the thiobios) may be a relict of very early metazoan evolution. The sand surrounding macrofaunal burrows is often densely colonized by meiofauna (Reise, 1981). This abundance is associated with high bacterial numbers and nutrient flux and thus may be linked to 'gardening'. Some meiofaunal species also garden, for example some nematodes produce mucus strings which are ingested after they have been colonized by bacteria. The meiofaunal habitat, besides being augmented by burrows, can also be extended above the sediment by colonization of projecting worm tubes (e.g. *Diopatra* and *Lanice*). There are some meiofaunal genera with representatives occurring in marine sands but also in subterranean fresh water which to this extent can be regarded as an extension of the interstitial habitat inland. The most comprehensive investigations of sand-beach meiofauna made to date have been in South Africa (see McLachlan and Erasmus, 1983).

Meiofaunal and benthic productivity

Meiofaunal standing crops in sand are often an order of magnitude lower than in mud. The latter values approach those for the macrofauna. There are still rather few published values, but meiofaunal biomass in shelf sediments is in the order of 50 mg l^{-1} compared with 200 mg l^{-1} of sand for macrofauna. However, due to its much more rapid turnover the secondary production of meiofauna can equal or exceed that of macrofauna (Gerlach, 1978). On an estuarine mudflat in England, where small polychaetes (e.g. *Manayunkia*), nematodes and copepods were abundant, the meiofaunal production reached 20 g C m^{-2} y^{-1} whereas the macrofaunal production was about 5.5 g C (Warwick *et al.*, 1979). Mud-sediment copepods are generally more robust than sand-dwelling species and form an important food resource for young fish. Macrofaunal production values vary considerably from one community to another; in Long Island Sound at depths of 6–31 m it is about 30 g C m^{-2} y^{-1} but in some North Sea areas is only 1.7 g C.

The secondary productivity of shelf benthos relies on food from two main sources, namely coastal macrovegetation (on average each m^2 of shelf receiving 100 g C y^{-1} from this source) and phytoplankton (which contributes a similar amount). The latter input is partly from settlement, partly from inefficient plankton herbivory and partly from zooplankton faeces. The two latter routes may dominate during periods of thermal stratification.

Other aspects

There has been a recent vogue for cage-experiments in benthic ecology. Caged sediments nearly always develop greater biotic density and diversity (for example see Hulberg and Oliver, 1980). This is assumed to be due to the exclusion of predation by fish and macro-invertebrates, although some physical changes may also be involved. It is apparent from these and other observations that the role of predation in structuring benthic communities has been underestimated.

There is a general need for information on temporal changes, whether stochastic, seasonal or long-term. For example storms can produce massive mortality in benthic communities and lead to replacement of the original dominant species by others; it is not known how long such changes persist. Present understanding of benthic dynamics is also hampered by lack of information about the flux of material between the water column and sediments.

CHAPTER FIVE

BRACKISH-WATER ENVIRONMENTS

Whereas the intertidal zone can perhaps be envisaged as the interface between marine and terrestrial environments, estuaries are transitional between open marine and fully freshwater environments. Estuaries can be defined as semi-enclosed bodies of coastal water which retain a free connection with the open sea and within which sea water is measurably mixed with fresh water of terrestrial origin (Pritchard, 1967).

In geological terms estuaries are relatively recent, the majority of them having arisen since the last Ice Age. They are also geologically ephemeral because the constant deposition of sediment and its subsequent colonization and stabilization by various emergent plants results in a gradual infill of the estuarine system (see also Chapter 7). Estuaries vary in size from little more than tidal creeks to the lower reaches of large rivers such as the Amazon and Mississippi whose influences can be detected far out to sea. The variability in physical, chemical and biological properties of estuaries has profound effects on the nature of the estuarine biota.

Types of estuary

Estuaries have been classified into four major types according to their geological origin.

(1) *Coastal plain estuaries* (= drowned river valleys or rias). These were formed at the end of the Ice Age as the rising seawater level invaded low-lying coastal river valleys. They conform most closely to the classical concept of an estuary. They are also by far the commonest type and the ones on which we shall subsequently concentrate (e.g. Chesapeake Bay; R. Thames).

(2) *Bar-built estuaries* (= semi-enclosed bays or lagoons). These shallow estuaries occur principally in low-lying regions where shingle or sand deposited in bars running parallel to the shoreline partially isolates the

river water from the sea (e.g. Pamlico Sound in North Carolina; Dutch Waddenzee).

(3) *Tectonic estuaries*. These comprise a small group of estuaries in which the sea has invaded the land, not through a rise in sea-level but by tectonic processes often resulting from land subsidence, especially along fault lines (e.g. San Francisco Bay).

(4) *Fjords* (= drowned glacial valleys). These deep and narrow estuaries typically have a shallow submerged sill (accumulations of glacial deposits) at their mouth; this can greatly restrict water circulation at the deeper levels. Fjords are especially prominent along the coasts of Norway, Chile and British Columbia.

Major environmental features

Water circulation and salinity

These two factors can be considered together since they are closely interrelated. The basic pattern of estuarine flow is one in which fresh water flowing seaward at the surface is replaced by denser saline water flowing up the estuary at depth. These two layers of water, however, show varying degrees of mixing according to local topography, tidal velocity, the relative volumes of fresh water and salt water, and friction at their interface. Several basic patterns of water circulation have been recognized although these are essentially points on what is effectively a continuum of estuary types.

In the *highly stratified or salt wedge estuary* (Figure 5.1) sea water invades the river bed in the form of a distinct salt wedge. Less dense river water flows over the salt wedge and spreads seawards as a progressively thinner layer. Frictional forces at the interface cause some entrainment of sea water upwards into the freshwater layer and in order to compensate for this continual loss there is a slow upstream flow within the salt wedge. The surface water becomes progressively more saline as it moves seaward, but in the absence of any significant downward mixing of fresh water, bottom salinities are essentially those of the open sea. As mixing is limited, a marked discontinuity in salinity (= halocline) exists between the two circulating water masses. Salt wedge estuaries occur where river flow is large relative to tidal flow and where the width/depth ratio of the estuary is relatively small, for example, the Mississippi, which is 30 m deep in places.

Where tidal flow increases relative to river flow, two-way vertical mixing effectively destroys the salt wedge and a *moderately stratified or partially*

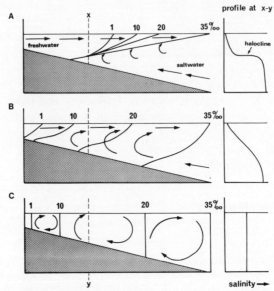

Figure 5.1 Distribution of isohalines in (*A*) a salt-wedge estuary; (*B*) a partially mixed estuary; and (*C*) a vertically homogeneous estuary.

mixed estuary (e.g. Chesapeake Bay) is generated (Figure 5.1). Salinity at all depths increases seawards, but at any given point the salinity of deeper water exceeds that at the surface. A larger volume of water is brought to the surface than in the salt wedge estuary. This is balanced by a similar volume of sea water moving landwards near the bottom. Even though mixing occurs there is still a degree of vertical stratification and a weak halocline persists. Where there is especially vigorous tidal mixing and relatively weak river run-off, all stratification is broken down resulting in a *vertically homogeneous estuary* (Figure 5.1). At any point in the estuary salinity is constant with depth (i.e. there is no halocline), though a marked salinity gradient extends along the length of the estuary. The Thames estuary shows the essential features of a vertically homogeneous estuary.

Superimposed on the above circulation patterns are those stemming from the Coriolis effect which deflects the water to one side of the estuary. Opposite banks of an estuary may thus exhibit markedly different salinity regimes. In extreme cases, especially in wide estuaries, the fresh water/sea water interface may be almost vertical.

The position of the mixing zone between the two water masses within the estuary will obviously change not only with the ebb and flow of the tide but also seasonally, depending on the volume of the freshwater input. In estuaries like the Amazon, where the freshwater discharge is exceptionally large, the mixing zone occurs out at sea. In especially hot climates a negligible input of fresh water may be offset by high evaporation at the surface. This dense hypersaline surface water then sinks and moves out of the estuary as a bottom current. Such estuaries have been called *negative estuaries*.

Sedimentation

Sedimentation processes are extremely complex (see Chapter 1 and Dyer, 1973, for details) and depend on the interplay of several factors including water movement, topography and the amount and type of available sediments.

Fine particulate material transported into estuaries by rivers undergoes a change in surface charge and flocculates when it encounters salt water (Wangersky, 1977). Larger floccules sink and form a liquid mud on the surface of the submerged mud-flats. When vertical mixing is vigorous some particles will be resuspended and carried back up into the freshwater layer. Here they deflocculate and the whole process is then repeated. Resuspension of particulate material is also facilitated by the biogenic activity within the sediment itself.

Not all the particulate matter reaching the estuary is of freshwater origin. Large amounts of sediment carried in by the tidal flow are deposited wherever water movement is sufficiently reduced. Estuaries are therefore efficient sinks for fine muddy sediments. Whilst sedimentation is broadly related to the seasonal input of fresh water, storms and floods can dump large amounts of sediment into estuaries on a very irregular basis. Most estuaries are muddy, but where rivers drain hard igneous rock or where currents are sufficiently strong to prevent the settlement of fine particles, sandy estuaries may develop.

Detritus or POM also settles out in large amounts in estuaries. Some of this is imported into the estuary from other sources (allochthonous) though much of it originates from within the estuary itself (autochthonous). The high productivity frequently associated with the estuarine ecosystem is largely due to the detrital food chain (p. 87).

Other factors

Estuaries frequently experience a much wider range of *temperature* than do adjacent coastal waters. This is due in part to their relatively large surface area/volume ratios and in part to the variable input of fresh water which is more subject to seasonal temperature changes. Estuaries are usually regions of calm water where *wave action* is minimal. However, in certain funnel-shaped estuaries (e.g. R. Severn) the rising tide is accompanied by a rapidly advancing wave of water. These 'bores' are often several feet in height and their effects felt far inland. Because of the high organic content and high bacterial populations in estuaries *oxygen levels* are generally low. Sediments in particular become severely depleted of oxygen and are often anoxic below the immediate surface layer. In fjords there may be severe deoxygenation of water below the sill. Most estuaries experience high levels of *turbidity*, especially during the periods of maximum river flow. This markedly reduces the amount of light penetration with consequent implications for many photosynthetic plants. *Currents* are generated by tidal and river flow and lead to channel formation and local erosion of mud-banks. Since sea water and fresh water differ with respect to their ionic content the *chemical composition* of estuarine water will vary according to the water budget of the particular estuary (Barnes, 1984).

The biota of estuaries

Estuaries are highly variable, unstable and stressful environments (especially with regard to salinity) and it is perhaps not surprising that relatively few species have evolved the necessary adaptations required to live there. Species richness is therefore generally much lower than in adjacent marine and freshwater environments. Some groups (e.g. cephalopods, echinoderms) are completely absent. However, those species which have successfully invaded brackish waters (e.g. *Macoma baltica*) often occur in habitats which they do not normally occupy elsewhere (= niche expansion) and at exceedingly high population densities (and/or biomass), trends which presumably reflect reduced levels of interspecific competition and an abundance of food. The relatively recent origin of most estuaries coupled with their general lack of habitat diversity (most estuaries are uniformly muddy) also probably contribute to the rather depauperate nature of the estuarine biota.

Because of the stratified nature of many estuaries, benthic organisms can penetrate the estuary further than can pelagic forms. Organisms in the mid-upper shore encounter more or less fully saline water when covered by the tide and are exposed to air at low tide. Organisms further down the shore experience greater extremes of salinity than those further upshore since they are covered by fresh water at low tide. This often permits marine animals to penetrate further up an estuary in the higher intertidal zone. Moreover, infaunal species experience less salinity variation than do epifaunal species because of the buffering effects conferred by the interstitial water which is less variable than the overlying water mass.

Three major faunal groups can be recognized; marine, freshwater and brackish-water species (Figure 5.2). Of these the marine component is the largest and includes both stenohaline and euryhaline species. Stenohaline marine species tolerate only a narrow range of salinity change and are restricted to the lower reaches of the estuary (e.g. *Cerastoderma edule*); euryhaline species tolerate wider salinity fluctuations (some down to salinities as low as 5‰) and consequently extend further up the estuary (e.g. *Carcinus maenas*). Freshwater species rarely tolerate salinities > 5‰ and occur principally at the head of the estuary. Salinities of 5–8‰ therefore appear to represent a *critical salinity range* for many species and accordingly mark the area of minimum species richness. A small group of brackish-water forms (e.g. *Hydrobia, Nereis diversicolor, Macoma*) do, however, occur in the middle reaches of the estuary where salinity fluctuations are especially pronounced. These are of marine origin but their seaward distribution appears to be restricted primarily by biological

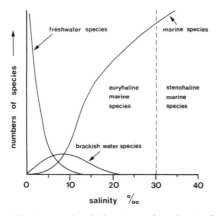

Figure 5.2 Numbers of freshwater, brackish-water and marine species along the estuarine gradient. (After Remane and Schlieper, 1971.)

interactions (predation, competition) with other species. Whether they are truly 'estuarine' is perhaps questionable since they occur in suitable fully saline habitats.

Apart from the resident fauna of estuaries, dominated by polychaetes, lamellibranchs and various crustaceans (especially amphipods and iso-pods) in cooler waters, and by gastropods and crabs in the tropics and subtropics, numerous migratory species pass through estuaries *en route* to breeding grounds in rivers or in the sea (see Chapter 9). Other migrants, including fish, birds and decapod crustaceans, use estuaries as feeding and/or nursery grounds.

Estuarine vegetation is also rather impoverished. Muddy sediments are generally unsuitable for the attachment of macroalgae though some species tolerant of reduced salinity (e.g. *Enteromorpha, Ulva*) may be seasonally abundant. High turbidity restricts light penetration and severely limits the growth of phytoplankton. In the lower reaches of estuaries sea-grasses may occur. However, by far the most important components of the estuarine macroflora are the emergent rooted plants. Because of their high organic content estuarine muds are rich in micro-organisms and meiofaunal species. The diversity of these stands in marked contrast to the paucity of larger organisms. For further details of these and other muddy-shore biota see Chapters 4 and 7.

Adaptations of estuarine organisms

Estuarine organisms face special and difficult conditions. In small tidally-mixed estuaries salinity gradients may change rapidly from fully freshwater to fully marine conditions over a single tidal cycle, allowing little time for acclimation. Salinity gradients in larger estuaries change more gradually. Even so they may be subject to marked seasonal freshwater inputs.

Mechanisms of adaptation to changing osmotic and ionic gradients vary considerably. Some species are *osmoconformers*, regulating neither cell volume nor ionic composition. The ability of these species to penetrate estuaries depends largely on their tolerance of wide variations in the concentration of their body fluids. Most marine species, however, exhibit varying degrees of *osmoregulation* and *ionic regulation*. In some species, including those incapable of regulating their extracellular fluids, some intracellular regulation may be achieved by altering the concentration of dissolved free amino acids in response to changes in external salinity. In this way intra- and extracellular fluids remain more or less isosmotic and a constant cell volume is maintained. For details of animal osmoregulation, see Rankin and Davenport (1981).

Many higher crustaceans are powerful osmoregulators, thus accounting for their successful exploitation of the estuarine environment. Most bivalves are osmoconformers and close their shells when salinity falls below acceptable levels. *Nereis diversicolor* osmoregulates only at lower salinities; above 10–15‰ its internal fluids conform to those of the external medium (Figure 5.3). By contrast, the mud prawn *Upogebia africana* maintains normal salinity levels in its tissues even when the salinity of the surrounding water drops to 2‰ (Branch and Branch, 1983).

Osmoregulatory efficiency appears to be correlated with temperature. This may partially explain why there are more estuarine species in the tropics and why stenohaline species extend further upstream than they do in estuaries at higher latitudes. It may also partly explain the seasonal migrations of some species (e.g. *Carcinus maenas*, flounder) into deeper waters during the cooler parts of the year.

Many estuarine organisms avoid the worst rigours of salinity changes either by burrowing into the sediment or by migrating. Some species, like the shore crab *Carcinus maenas*, move up and down the estuary with the tide in order to maintain themselves in a relatively restricted salinity range. Many crustaceans move out of estuaries to breed, since their eggs and larvae are more sensitive to low salinity than the adult (e.g. the blue crab *Callinectes*). Conversely, other species (e.g. mullet, flounder) use estuaries

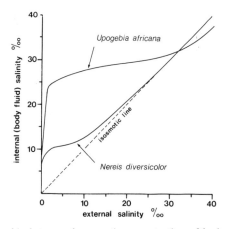

Figure 5.3 Relationship between the osmotic concentration of body fluids and ambient salinity in two estuarine organisms.

as nursery grounds. Estuarine circulation can potentially result in the loss of many planktonic larvae. Many species are therefore adapted to remain within the estuary by selectively swimming into the water column on the flood tide and remaining in or near the bottom on the ebb (e.g. spionid polychaetes). Many estuarine invertebrates have effectively eliminated the planktonic stages from their life-cycles. The larvae of *Nereis diversicolor* for instance are benthic whilst the eggs of the sand prawn *Callianassa kraussi* hatch directly as miniature adults. In other species (e.g. the mud-snail *Nassarius kraussianus*) the planktonic stage is kept suitably brief.

Certain morphological/anatomical differences (e.g. smaller maximum size, reduced number of vertebrae in fish) have also been demonstrated in estuarine populations. Many of the other structural specializations are essentially adaptations for living in muddy sediments (see Chapter 4). Reproductive rates of marine species are often lower in brackish conditions, whilst freshwater species may be partially sterile. It would appear, therefore, that many estuarine populations have diverged into genetically distinct races relative to conspecific populations on adjacent open coasts.

Productivity and food webs

Estuaries can be exceedingly productive, comparable in this regard to tropical rain forests, coral reefs and kelp beds. Phytoplankton, benthic algae and rooted plants all contribute to primary production in estuaries. However, because of high turbidity and distinctive patterns of water circulation (in which there is a net transport of surface water to the open sea) phytoplankton are clearly at a disadvantage in many estuaries. Moreover, macrophytic plants compete with phytoplankton for nutrients and some (e.g. mangroves) may even release tannins which effectively suppress phytoplankton growth. Nonetheless, phytoplankton production can be substantial, especially in larger estuaries where the interchange with the open sea is restricted. On the whole, estuarine zooplankton species are inefficient grazers of phytoplankton and some 50% of the net phytoplankton production may thus be available to benthic forms either as living cells or ultimately as detritus. Most seaweeds are relatively intolerant of low salinities and are thus usually confined to the lower reaches of those estuaries with extensive connections to the open sea. Diatoms living on or within the surface of the mud often provide an important food source. A large proportion of the primary production in many estuaries, however, comes from the rooted plants which grow on the intertidal mud-banks and around the margins of the estuary. As in many marine macrophytic communities very few herbivores feed directly on the vegetation. Much of this production therefore eventually circulates through the detrital pathway. It is this autochthonous detritus together with organic matter washed in by rivers and tides that ultimately provides the major food resource for estuarine animals and thus fuels this ecosystem. Precisely how much detritus is contributed from each source is difficult to ascertain. In some estuaries, marsh grasses or mangrove leaves appear to be the major contributors to the detrital pool; in others planktonic and benthic algae appear more important. Nutrients released from detritus and animal faeces after decomposition processes can be moved back upstream in the bottom water and so become available for further plant growth.

In view of the enormous importance attached to the detrital food chain (Chapter 4) it comes as no surprise that estuaries are rich in detrital feeders. These are mainly benthic invertebrates and include both deposit feeders and suspension feeders. These two methods of feeding, however, are by no means mutually exclusive. Deposit feeders are especially diverse in fine estuarine sediments. Suspension feeders are rarer, presumably because their delicate filtering devices are easily clogged. Some, however, including several commercially exploited species (e.g. *Mytilus edulis, Crassostrea virginica*) may be locally abundant. The assimilation efficiency of many detritus feeders is low and sediments are therefore continually being reworked (= bioturbation). Nonetheless, the plentiful supply of food enables detrital feeders in estuaries to attain exceedingly high population densities and biomass.

Detrital feeders are consumed either directly or indirectly by various predators. Some crabs, birds and fish are among the most important. As in other ecosystems predators have a marked effect on community structure.

The major interactions in a typical estuarine food web are shown in Figure 5.4. It is the ability of estuaries to function as efficient nutrient traps, together with the extensive recycling of these nutrients between the water and the biologically active bottom sediments, that largely underlies the high levels of production associated with estuarine ecosystems. The seasonal abundance of food makes estuaries especially favourable as nursery areas for fish and various invertebrates and as feeding grounds for migrant birds. Perhaps one of the most controversial areas of estuarine ecology is the extent to which estuaries export organic material and thereby subsidize production in adjacent coastal waters. Available evidence is inconclusive but suggests that nutrient outwelling may be important in some estuaries but not in others. Outwelling may be periodically accelerated as a result of coastal upwelling (Duxbury, 1979).

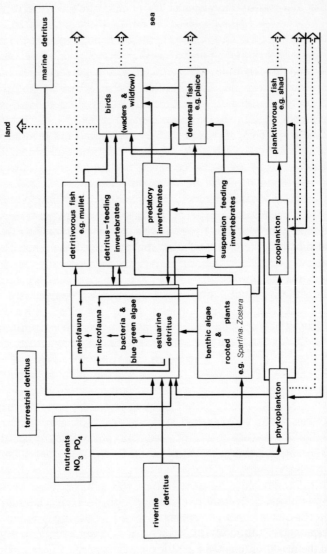

Figure 5.4 Generalized estuarine food web. Dotted lines denote 'losses' from the estuarine system (After Barnes, 1984.)

Brackish and hypersaline environments

Estuarine realms occur beyond river mouths where large volumes of freshwater discharge result in areas of coastal water in which the surface layer is distinctly brackish. They are bounded on the landward side by the coastline and on the seaward side by a salinity front. Waters off the eastern coastline of North America are a typical example. *Hypersaline lagoons* are shallow coastal bodies with narrow restricted inlets to the sea and little or no freshwater input. They occur particularly where there is a longshore or onshore movement of sediment. In the absence of any significant freshwater input or strong tidal exchange, water circulation is generally very sluggish. Evaporation results in high (though relatively stable) salinities and this together with extremes of temperature, high pH and the accumulation of organic sediments results in a quite distinctive biota which may be extremely productive. Hypersaline lagoons occur mainly in the tropics and subtropics (e.g. Bitter Lakes, Egypt, and the Sivash, USSR).

Brackish seas (e.g. Baltic) are large semi-enclosed bodies of water in which salinity may be lower than in many estuaries. They are distinct from estuaries, however, in that their waters are *permanently* brackish. The osmotic problems facing organisms in this environment thus differ somewhat from those in estuaries. The brackish seas of the USSR (e.g. Caspian) have a long and complex history. Consequently their biotas are diverse and include both freshwater and relict marine components.

Tidepools on rocky shores would appear to provide suitable habitat for many marine organisms. However, physical conditions, particularly in smaller mid- to high-shore pools, fluctuate widely and consequently many common species which lack the appropriate physiological specializations are absent. Salinity varies according to rainfall and evaporation during low-tide exposure. Temperature, oxygen and pH also vary tidally. Larger low-shore pools are largely buffered against such fluctuations and therefore correspond more closely to a subtidal habitat. Accordingly these often have a more diverse biota. The absence of some species from tidepools may be due to competition from those species more perfectly adapted to pool life.

Brine seeps occur subtidally in some areas where hypersaline water emerges on the sea-floor (e.g. the East Flowergarden seep in Texas). These are characterized by mats of anaerobic prokaryotes with a few associated meiofaunal species.

CHAPTER SIX

CORAL REEFS

For their sheer beauty, species richness and multiplicity of biological interactions, coral reefs have few equals in the natural world. They are also highly productive systems in waters which are otherwise low in nutrients and phytoplankton. Their complex, wave-resistant structures are built almost entirely by biological activity. The dominant reef-forming organisms are the scleractinian (= stony) corals which grow by means of an accretionary exoskeleton, but the lithifying (cementing) properties of crustose coralline algae, aided by the precipitation of magnesium carbonate particles from the water, are also very important in maintaining the structural integrity of the reef framework, especially in shallow waters.

Most of the structure of a coral reef consists of dead skeletal limestone deposited over years of accretion; living coral tissue forms only a thin veneer of interconnecting polyps over the surface of this extensive matrix. In addition to stony corals and coralline algae many other calcareous algae (e.g. *Halimeda*) and lime-secreting invertebrates (e.g. sponges, hydrocorals, octocorals, molluscs) also contribute to the formation of the reef.

Corals occur throughout the oceans, but colonial reef-building (hermatypic) corals are confined to the tropics and subtropics where calcification rates are greatly enhanced. *Lophophelia* is a branching coral found in cooler waters. Solitary corals do occur on coral reefs but these tend to be much larger than their counterparts from higher latitudes. One of the characteristic features of hermatypic corals is the presence of unicellular algae (*Symbiodinium*) within the gastrodermal tissue. By their influence on coral growth and calcification rates these symbiotic zooxanthellae play a fundamental role in the reef-building process.

Classification

Three major geomorphological types of coral reefs can be recognized—fringing reefs, barrier reefs and atolls (Figure 6.1). Fringing

90

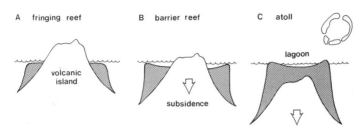

Figure 6.1 Atoll formation according to Darwin's Subsidence Theory.

and barrier reefs, however, are not easily separated from each other and are sometimes grouped together as *coastal reefs*, since both occur adjacent to land masses which they effectively protect against the damaging effects of waves. Coastal reefs are common throughout the coral-reef zones in all oceans whereas atolls occur mainly in the Indo-Pacific region.

Fringing reefs grow seawards, leaving an ever-widening reef flat between the actively growing edge of the reef and the shoreline. They are the commonest type of reef and occur for example in the Red Sea and along the northern coast of Jamaica. *Barrier reefs*, like fringing reefs, run parallel to the coastline but are separated from the land mass by a shallow lagoon. The Great Barrier Reef (actually a collection of reefs) off the Queensland coast of Australia is almost 2000 km long and covers an area of more than 207 000 km^2. *Atolls* are crescent or ring-shaped reefs surrounding a central lagoon. Parts of the reef platform may emerge as an array of small islands. Atolls occur in deep water and are not associated with any obvious land mass. *Patch reefs* are small circular or irregular reefs that arise from the floor of lagoons behind barrier reefs or within atolls.

Darwin suggested that atolls were formed from fringing reefs which developed on the shores of slowly subsiding volcanic islands. As these islands sank, vigorous upward growth of coral at the periphery of the reef produced first a barrier type reef and then, as the island finally disappeared beneath the sea, an atoll. Continued growth along the outer edge keeps the reef at the surface but calmer conditions, sedimentation and bioerosion on the inside discourage vigorous growth and a lagoon develops. Darwin's 'Subsidence Theory' was subsequently confirmed in 1953 when corings at Eniwetok atoll in the Marshall Islands revealed volcanic rock beneath more than 1000 metres of reef limestone. Darwin's theory conveniently links all three reef types in an evolutionary sequence. However, it must not be assumed that all fringing or barrier reefs are gradually evolving into atolls. Indeed coastal reefs, unlike atolls, probably have very diverse origins

D

and histories, Atolls are exceedingly old (up to 60 million years) but many other reefs are much more recent, some dating back only to the last Ice Age (< 15 000 years).

Distribution and limiting factors

Coral reefs are widely distributed wherever there is a suitable climate and substratum on which they can grow. They occur principally within the 20° isotherm (Figure 6.2) but attain their maximum development where the mean annual water *temperature* is 23–25°C. No significant reef development occurs below 18°C, though higher temperatures (up to 40°C) can be tolerated. The absence of reefs along much of the western coastlines of South America and Africa is due to the strong upwelling of cold water. Occasionally, as in the western Atlantic, warm surface currents enable coral reefs to extend into higher latitudes. Recent evidence (Johannes *et al.*, 1983) suggests that coral distribution may be set through the interaction of temperature and competition with macroalgae. Lower temperatures which are suboptimal for corals appear to favour macroalgal growth; thus competition and not temperature *per se* may effectively exclude corals from latitudes where calcification rates are still potentially high.

Adequate *light* is essential for the photosynthetic activity of the zooxanthellae which permeate the coral tissue. Since light intensity decreases exponentially with depth (see Chapter 2) active reef building rarely occurs below about 20–30 m. The compensation depth for coral growth occurs where the light intensity is approximately 15–20% of the surface intensity. Coral reefs are thus largely restricted to the margins of

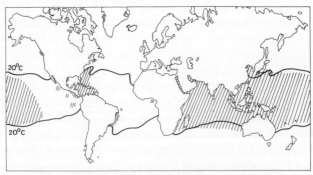

Figure 6.2 World distribution of major coral reef areas.

continental land masses and oceanic islands. Some corals, however, are adapted to lower light intensities and soon die if exposed to high surface illumination.

Coral reefs are very sensitive to low *salinity* and *sedimentation*. Consequently they are absent wherever there is an excessive influx of silt-laden fresh water even though conditions might otherwise be favourable. The absence of reefs along much of the tropical coastline of the Atlantic coast of South America for example is due to the influence of large rivers such as the Amazon and Orinoco. Optimal conditions occur in the Red Sea, around offshore islands in the Caribbean and Indo-Pacific regions and along the east coast of Australia, areas where terrestrial run-off is minimal or locally restricted. Coral reefs are tolerant of hypersaline conditions and in the Persian Gulf flourish at salinities exceeding 40‰. Turbidity reduces light penetration whilst settling silt particles smother the feeding structures of the delicate coral polyps. Some corals, however, especially those with larger polyps, produce moving sheets of mucus which effectively carry any sediment off the surface of the coral; these species are therefore more tolerant of turbid conditions.

Coral reefs thrive in regions of strong water movement. *Wave action* prevents sedimentation and keeps the water well oxygenated. Although the rigid calcareous skeletons of most hermatypic corals are particularly resistant to wave-shock, severe storms do occasionally cause extensive damage and are thus a major source of change in many coral-reef communities (see p. 102). Corals are unable to withstand prolonged *aerial exposure* and rarely grow much above the low spring-tide level.

Reef morphology and zonation

In response to steep gradients in the physical environment, coral reefs, like rocky shores, are conspicuously zoned as a series of broadly overlapping bands running roughly parallel to the shore. Reef morphology is exceedingly complex, but whilst detailed structure and composition varies considerably from one locality to another the basic pattern in all types of reefs is similar. The pattern for atolls, although generally more complex than other reef systems, will serve as an example.

Basically an atoll consists of a sheltered central lagoon separating windward and leeward reefs on which there may be one or more low islands formed from consolidated coral debris (Figure 6.3). The windward *reef front* is characterized by vigorous coral growth in the form of massive

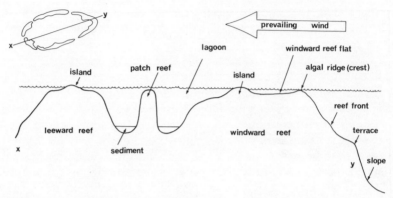

Figure 6.3   Diagrammatic section of a typical atoll showing the major subdivisions of the reef complex.

finger-like spurs or buttresses running normal to the reef edge and extending down to the *reef terrace* at a depth of about 15–18 m. These spurs, which may be up to 300 m long depending on the slope of the reef, alternate with deep grooves or surge channels. This spur and groove formation (Figure 6.4) effectively dissipates the destructive energy of the waves by channelling water up the grooves and over the seaward edge of the reef. As the water washes back down the channels it is loaded with sediment and coral debris from the reef above; this erodes the grooves still further. The spurs and walls of the channels are optimal sites for coral growth and it is in this region that maximum accretion occurs. Dominant reef-forming corals (e.g. *Acropora, Montastrea, Pocillopora, Porites*) are abundant in this buttress zone. Sometimes corals from adjacent spurs coalesce over the surge channels forming canyons or tunnels which open through distinctive 'blow holes' on the reef crest.

The lower limit of the windward reef front is marked by the reef terrace below which the *reef slope* extends down, sometimes vertically, to the sea-floor. Although living corals occur down to depths of around 50–70 m their growth is here much less prolific and the reef slope is typically colonized by more delicately branching and plate-like forms. These corals (e.g. *Agaricia*) have large surface areas and are thus well adapted for intercepting the reduced amounts of light that penetrate to these deeper regions. Even individual species (e.g. *Montastrea annularis*) are known to change their morphology in response to differing light levels. Moreover, flattened plate-like forms are also less likely to roll down the reef slope to lethal depths should they become dislodged by waves or undermined by the biological

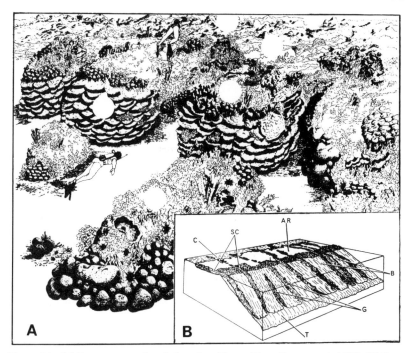

Figure 6.4 (*A*) Buttress zone of a windward reef front. (From Stearn *et al.*, 1977). (*B*) Spur and groove formations showing surge channels (*SC*), terrace (*T*) algal ridge (*AR*), buttresses or spurs (*B*), grooves (*G*) and coral of reef flat (*C*). (After Yonge, 1963.)

activity (= bioerosion) of various reef boring organisms (e.g. fungi, sponges, bivalves). Lower down the reef calcareous sponges, octocorals and non-hermatypic stony corals become increasingly important.

Immediately behind the buttress zone the seaward crest is dominated by crustose coralline algae (e.g. *Lithothamnion, Porolithon*). These form a smooth *algal ridge* which receives the full impact of the breaking waves. The crest is dissected by smooth-sided surge channels. Algal ridges are widespread on Indo-Pacific reefs but rarely occur on Atlantic reefs. A rich invertebrate fauna including crabs, shrimps and gastropods finds refuge in the subsurface cavities of the ridge where it is effectively protected from waves and predators. Zoanthids (anemone-like anthozoans) and vermetid gastropods may also be abundant in this zone whilst various urchins (e.g. *Diadema*) 'burrow' into the calcareous surface. Where wave action is less severe, reef crests may be dominated by robust corals such as *Millepora* (a hydrocoral) and *Acropora* whose heavy spreading branches form a seaward-projecting thicket.

Behind the algal ridge is the broad *reef flat* (back reef) which is very variable in character. Changes in salinity, temperature and sedimentation are most extreme in this part of the reef. This results in a large number of different habitat types and this zone is often variously subdivided. Reef rubble dislodged by waves sometimes forms a well-defined boulder zone at the outer edge of the reef flat. Conditions for coral growth are most favourable near the algal ridge where waves spill over the reef crest. Elsewhere on the reef flat reduced water circulation, accumulation of sediment and tidal exposure prevent extensive coral growth, though some massive species (e.g. *Porites*) may occur, especially in the deeper regions. Hard surfaces are frequently colonized by calcareous algae and unconsolidated sediments by sea-grasses, each community supporting a rich assemblage of invertebrates. In protected environments on the inner reef flats benthic foraminiferans may be abundant.

Inside the reef flat is the *lagoon*. This is rarely more than 60 m deep and the floor is typically covered with sediments derived largely from erosion of the windward reef. These sediments support numerous infaunal species whilst extensive beds of sea-grasses (e.g. *Thalassia*) and calcareous algae (e.g. *Halimeda*) may also be present. Reduced water movement and increased sediment create conditions which are generally unfavourable for extensive coral growth, though reefs varying in size from small knolls to larger patch reefs do occasionally rise from the lagoon floor wherever suitable conditions prevail.

The leeward reef is essentially similar to the windward reef though it is narrower and generally lacks the algal ridge, buttresses and surge channels. Extensive coral growth occurs at the seaward edge where, in addition to massive species, faster-growing branching species occur in the less wave-swept conditions. This region is rich in invertebrates (especially urchins) and fish.

Whilst lacking the complexity of atolls, the morphology and zonation of coastal reefs along windward and leeward shores of land masses is essentially similar to that described respectively for windward and leeward reefs of atolls. In barrier reefs the inner margin of the lagoon is of course the coastline itself.

Nutrition, calcification and growth

Corals are carnivores and use their nematocyst-laden tentacles to capture zooplankton. Although individual polyps are small they are exceedingly numerous and a coral colony thus presents an extensive feeding surface.

Ciliary mucus-traps and extrusion of mesenterial filaments through the mouth provide additional plankton-capturing mechanisms. Despite the general paucity of plankton in the surrounding tropical oceans, coral reefs are now known to support abundant zooplankton populations. Much of this is indigenous and quite distinct from that in the surrounding water (Sale *et al.*, 1978). This plankton consists largely of demersal forms which emerge from the bottom and reef interstices only at night. Even so, recent work has shown that the zooplankton probably supplies only a small fraction (5–20%) of the daily energy requirements of most corals, though it may be a more important component in the diets of some particularly voracious carnivores.

The use of radioisotopes has shown that photosynthates (glycerol, glucose, alanine) produced by the zooxanthellae are utilized directly by the coral. In some species these photosynthates can potentially provide for most of the carbon required by the coral. However, unlike many other organisms which also harbour symbiotic algae (e.g. tridacnid clams) corals do not digest their zooxanthellae even when starved. Some corals can survive and grow for several months using only the photosynthetic exudates of their algal symbionts. A continuum of coral types exists, ranging from the most voracious species at one extreme to those which depend largely on symbiotic photosynthesis at the other. Typically, carnivorous species have large polyps for capturing mobile prey and form massive colonies with low surface/volume ratios; specialized autotrophs, by contrast, have smaller polyps and foliose or plate-like growth forms (i.e. high s/v ratios) for the more efficient interception of light (Porter, 1976). Corals which are essentially autotrophic expand their polyps by day whilst carnivorous species feed nocturnally on plankton. Most carnivores, however, still rely heavily on their zooxanthellae as an energy source. Corals may also obtain some of their nutrient supply from bacteria and from the active uptake of DOM (see Muscatine and Porter, 1977).

Zooxanthellae also promote calcification, though the precise mechanisms involved are not fully understood. One possible mechanism is illustrated in Figure 6.5. Here calcium ions are actively transported through the gastrodermis and combined with bicarbonate ions to form first soluble calcium bicarbonate and then crystals of calcium carbonate which are incorporated into the exoskeleton. Zooxanthellae may be involved in this process since they utilize carbon dioxide for photosynthesis, thus shifting the equilibrium from bicarbonate to carbonate. A difficulty with this hypothesis is that sea water is supersaturated with calcium carbonate, and removal of carbon dioxide by the symbionts is

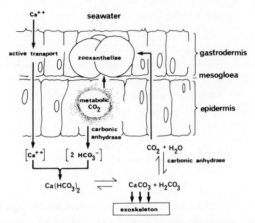

Figure 6.5   Calcification in corals. (After Yonge, 1963.)

perhaps unlikely to have any appreciable effect on the rate of calcification. Moreover it does not explain why the apical polyps of some corals calcify faster than lateral polyps even though they contain fewer symbionts. Alternative hypotheses have suggested that zooxanthellae stimulate calcification by (1) providing organic material for the synthesis of the skeletal matrix, (2) removing potentially toxic metabolites from the calcifying micro-environment, and (3) providing glycollate which is subsequently involved (as allantoin) in the transport of calcium and carbon dioxide to the sites of calcification.

Coral colonies grow by asexual proliferation, the rate varying according to their identity, age and habitat. Linear growth is faster in foliose and branching species although their final weight is often much less than that of massive forms. Some massive species are extremely long-lived (> 100 y) but most living reef corals are probably less than 10 years old. Growth of younger colonies is usually more vigorous and less variable than in older, larger colonies. The upward growth of whole reefs is reported to be 0.2–8.0 mm y^{-1}. Water movement and depth (through its effects on the amount of available light) are important determinants of coral growth and the form of individual species. *Montastrea annularis*, for example, occurs as a massive hemispherical form in the wave-swept surface waters (10 m) but is plate-like at greater depth (30 m). Growth is also affected by temperature and for any particular species is faster in warmer parts of its range. Seasonal variation in skeletal density produces distinct growth bands which can be used to determine age and growth rate. Corals where zooxanthellae have been experimentally removed show markedly reduced growth rates, thus emphasizing the important role of these endosymbionts in the reef ecosystem.

Species interactions

Mutualism

Symbiotic relationships are especially important in structuring coral-reef communities. One such relationship is that between corals and their zooxanthellae; although both can live independently, the relationship is mutually beneficial. Excretory wastes (carbon dioxide, ammonium phosphate) produced by the coral are removed by the alga and used in photosynthesis. In return oxygen and photosynthate are made available to the coral and calcification promoted. Zooxanthellae also occur in other reef organisms such as ascidians and sponges and in the mantle edge of tridacnid clams.

Crabs and shrimps form many commensal relationships. For example a symbiotic crab (*Trapezia*) and a pistol shrimp (*Alpheus*) actively defend their host (*Pocillopora*) from attack by the asteroid *Acanthaster* which then preys on other corals (Glynn, 1976). An interesting mutualistic relationship is that between large predatory fish and smaller cleaner-fish (e.g. *Labroides*). By removing ectoparasites and even dead or diseased tissue, cleaner-fish may be important in maintaining the general health of the predator. The bold colour patterns and characteristic undulatory movements of the cleaners serve as recognition stimuli and ensure that the cleaners are not eaten. During the cleaning process, which often takes place at specific 'stations', the larger fish assume non-aggressive postures. The cleaners in turn are mimicked by other fish which can thus approach and take bites out of the 'host' species. On Indo-Pacific reefs clown fish (e.g. *Amphiprion*) nestle amongst the stinging tentacles of large anemones. These small fish gradually protect themselves from the anemone's virulent nematocysts by coating themselves in anemone mucus, thereby mimicking their host. Stripped of this protective covering they are immediately killed and ingested. Their vivid colours and distinctive behaviour patterns attract other predatory species which are then more readily caught by the anemone.

Competition

As on many rocky shores, living space is often limited on coral reefs, and competition for this resource is potentially severe. Fast-growing branching corals are capable of overtopping slower-growing encrusting or massive species. Where this occurs the overgrown species no longer has access to an adequate light source and subsequently dies. Coexistence is facilitated however by the combined effects of physical disturbance and aggressive behaviour of the slower-growing species. Storms periodically break up branching corals but generally cause little damage to massive species. Slower-growing corals can extend digestive mesenterial (gastric) filaments either through the mouth or through openings in the body wall; these cause tissue necrosis when applied to adjacent colonies of other species (Lang, 1973). Corals can be arranged into *aggressive hierarchies* in which the slower-growing species are generally more aggressive than the faster-growing species. Some of the most aggressive species are small solitary corals which would otherwise be easily overgrown. The use of virulent sweeper tentacles and the production of toxic allelochemicals are additional mechanisms that prevent competitive exclusion of one species by another and thus serve to maintain diversity. Allelochemical aggression is also identified as an important competitive mechanism in many slow-growing reef sponges and ascidians. It has been argued that extensive coexistence is maintained by virtue of the 'non-hierarchal' nature of the competitive interactions where species A is superior to species B which in turn is superior to species C; however, by producing a poisonous toxin, species C is superior to A. This is termed a *competitive network* (Buss and Jackson, 1979). Reversals in competitive ability as environmental conditions favour first one species then another may also contribute to coexistence in these cryptic reef communities.

Predation

The importance of predators on coral reefs is less well documented than for rocky shores. Many small predators (e.g. gastropods, crabs) feed on coral polyps or coral mucus, but their overall effect on community structure is probably minimal. The two major groups of coral predators are starfish and fish. *Acanthaster planci* is a large starfish which feeds on live coral polyps. Normally it occurs at moderate densities and its presence may even promote diversity since it feeds selectively on competitively dominant species. However, since about 1960 populations of *Acanthaster* have increased dramatically and have caused extensive damage to many Indo-Pacific reefs. No-one knows for certain what has caused these population expansions. One suggestion is that the systematic removal by shell collectors of the giant triton *Charonia tritonis*, a natural predator of *Acanthaster*, may be partially responsible (Endean, 1977); other explanations include pollution and natural population fluctuations. Following destruction by *Acanthaster*, reefs often become colonized sequentially by filamentous algae, calcareous algae and octocorals. Several years may elapse before new corals become established and it may take up to 40 years for complete recovery.

A striking feature of coral reefs is the presence of myriads of brightly coloured fish which swim in and around the coral heads. The ecology of these fish is reviewed by Sale (1980). The precise mechanisms, such as niche segregation and random colonizations, which may allow so many fish species to coexist in abundance remain largely unresolved. Fish which are active on the reef by day differ from those which occur at night. Daytime fish are generally more brightly coloured and often occur in shoals. Shoaling behaviour enables approaching predators to be detected more effectively, whilst movements of the shoal confound them and permit members of the shoal to escape by darting for cover in reef crevices. Bold colour patterns serve several functions including species-recognition, camouflage and warning that the fish is toxic or unpalatable. By night these fish seek shelter within the reef and are replaced by a smaller number of nocturnal fish which feed principally on benthic invertebrates. Most coral-reef fish are predators. Some of those which feed on corals have highly modified teeth and jaws. Species may feed directly on coral polyps (e.g. pufferfish, filefish, triggerfish); others are more omnivorous and ingest living and dead coral fragments, digesting out the algae and endolithic (= boring) fauna (e.g. surgeonfish, parrotfish). The latter group are amongst the most common reef fish and are thus important producers of coral sediment. Whilst coral-reef predators generally cause little direct damage, bared coral skeleton is more accessible to various endolithic organisms whose presence weakens the coral colony. The widespread occurrence of antipredator defences (e.g. toxins) amongst reef organisms (e.g. Bakus, 1981) is testimony to the importance of predation within the reef ecosystem.

Grazing

Coral-reef algae are extensively grazed by various herbivores, especially urchins and fish, and marked changes in species composition and diversity of reef communities occur when these herbivores are experimentally excluded by caging. Moderate levels of grazing, however, appear to promote coexistence by preventing space monopolization by fast-growing algae. Very high densities of herbivores, particularly urchins, also lead to the removal of many recently established coral colonies with consequent implications for coral zonation (Sammarco, 1980). Most herbivorous fish are active by day. Many are territorial, having 'algal gardens' which they graze and vigorously defend. Some species (e.g. damselfish) even 'farm' their territories by actively removing unwanted species. Reef algae have evolved various defence mechanisms against herbivory—some for example are short-lived annuals, others live in relatively inaccessible habitats (crevices) and extensive use is made of structural (calcification) and chemical (tannins, ketones) defences.

Physical disturbance

Coral reefs often appear to be exceedingly stable and persistent systems. Nevertheless, they can suffer extensive destruction, particularly from the effects of tropical storms. Hurricanes and cyclones often overturn coral heads, destroy most branching forms and even disrupt spur and groove formations. Colonies are seldom completely obliterated however and reef recovery is facilitated by vegetative growth of surviving portions. It is estimated that some 25–30 years may be required for reefs to recover completely from the effects of particularly destructive hurricanes. Prolonged periods of tidal emersion, especially when these coincide with extreme temperatures or heavy rainfall, can also result in extensive damage. Delicate branching species are less resistant to desiccation than are massive corals.

   Although natural disturbances can be extremely damaging, moderate levels of disturbance may actually be beneficial to some corals by a process of thinning. The net effect is to open up space on the reef, thereby disrupting competitive exclusion and promoting coexistence. By transporting living fragments into new habitats storms may also serve as an effective means of coral dispersal. Coral reefs are thus strongly controlled by physical disturbance and in this regard are remarkably similar to many rocky intertidal communities (see Chapter 3).

Diversity gradients and biogeography

Coral reefs are extremely diverse, their complex three-dimensional structure providing suitable habitats for numerous invertebrates and fish. It is estimated that over 3000 species of invertebrates occur on certain reefs within the Great Barrier Reef complex. Polychaetes belonging to 103 species (13–1400 individuals) have been extracted from approximately 2.5 kg of dead coral rubble. One reason for this diversity is that the coral reef environment is a mosaic of different habitat types—wave-swept hard surfaces, sheltered sediments and carpets of algae and sea-grasses can often all be found within a relatively small area. Another reason is the relative stability of coral reefs over evolutionary time which has allowed ample opportunity for the coevolution of many specialized, finely-tuned interspecies relationships.

Two separate provinces of coral reef biotas can be recognized, one in the Indo-Pacific, the other in the Atlantic. The Atlantic province has significantly fewer coral species than the vast, yet relatively homogeneous Indo-Pacific province (62 species in 36 genera and 700 species in 80 genera respectively). Coralline algae are generally less important in Atlantic reefs where many invertebrates (e.g. tridacnid clams, giant sea anemones and their commensal fish) are also absent. Horny gorgonian corals (sea-whips and sea-fans), are particularly conspicuous on Atlantic reefs, whereas soft corals (alcyonaceans) are more prominent in the Indo-Pacific. *Acropora* is a dominant reef builder in both provinces but whilst it is represented by 150 species in the Indo-Pacific, only 3 species occur on Atlantic reefs. In both provinces the number of coral genera declines as one moves away from the centres of distribution (Figure 6.6). Since temperature is strongly correlated (positively) with coral diversity, the latter also declines more rapidly along the N–S than along the E–W axis, particularly in the Indo-Pacific.

Indo-Pacific reefs are much older than those in the Atlantic. The latter are generally situated on shallow platforms cut by wave action during the Pleistocene period when sea-level was much lower than it is today. Most Atlantic reefs are therefore of fairly recent origin dating back only to the last Ice Age. Evidence suggests that the rate of evolution in the Indo-Pacific may be approximately double that in the Atlantic; this, together with the greater age and environmental stability of Indo-Pacific reefs, probably

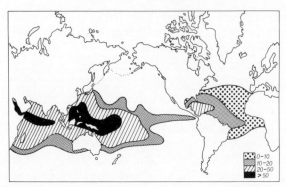

Figure 6.6 Generic diversity of reef-building corals in the Indo-Pacific and Atlantic biogeographical provinces. Key denotes number of genera. (After Stehli and Wells, 1971.)

accounts for the difference in species diversity between these two provinces. Some of this difference could also be accounted for by the much greater size of the Indo-Pacific (an 'area effect').

Maximum coral diversity occurs at depths of about 15–20 m where light conditions and water movement are most favourable. On some Indo-Pacific reef slopes, coral species gradually replace each other with increasing depth. On Caribbean reefs, on the other hand, fewer species are present over the depth gradient and many of these simply respond to decreasing light levels by altering their gross morphology (p. 94).

Productivity

Favourable light conditions, high temperatures and constant water movement (to replenish nutrients and enhance diffusion rates) provide ideal conditions for plant growth. These, together with the tight recycling of nutrients, combine to make coral reefs one of the most productive of all communities (see Lewis, 1977). However, whilst gross primary production may be $1500–3500 \, \mathrm{g \, C \, m^{-2} \, y^{-1}}$ (compared with around $18–50 \, \mathrm{g \, C \, m^{-2} \, y^{-1}}$ in the surrounding tropical waters) most of this production appears to be consumed within the community itself, strongly suggesting that reefs are virtually self-contained ecosystems.

Almost every conceivable type of marine primary producer contributes to this high level of production and although the total biomass of photosynthetic material is often relatively inconspicuous, on certain reefs it may actually exceed that of the corals themselves. Primary producers include zooxanthellae, filamentous green algae living on or within the coral skeleton, macrophytic red and brown algae, crustose and other calcareous algae, sea-grasses, benthic diatoms, phytoplankton and blue-green pro-karyotes, some of which occur symbiotically with reef sponges. No two reefs are alike and the precise contribution that each of these components makes to total primary production still remains to be determined, though present evidence strongly suggests that zooxanthellae (which can account for over 70% of the tissue weight of corals) are extremely important. Macro-algal turfs, which are extensively grazed by guilds of invertebrates and fish, and sea-grass beds which often occur in sheltered coral-reef sediments, are also identified as sites of intense primary production. Crustose algae, on the other hand, although important in reef lithification appear to be rather less important sources of production. Most sea-grass production eventually enters the detrital food chain (Chapter 4). The ability of sea-grass rhizomes and some prokaryotes (e.g. *Calothrix*) to fix atmospheric nitrogen provides the reef system with an important source of this nutrient in an otherwise nutrient-deficient environment.

Phytoplankton imported from the open sea is not a significant source of primary production in coral reefs, but together with incoming zooplankton and inorganic material it does provide a small but nonetheless important external source of nutrients which help compensate for nutrient loss from the reef ecosystem. Once incorporated into the food network these nutrients (especially phosphorus) are efficiently recycled within the reef. Instrumental in this recycling process are the symbiotic algae which are effectively captive within the system and cannot be washed away. Moreover, metabolic wastes produced by the host are immediately available to the algae, thus avoiding the loss of these valuable resources which would result from their release into the surrounding water.

Although coral reefs are extremely diverse communities by far the most important consumers of the primary production are the corals themselves. These, as we have already seen, feed partly on the resident reef zooplankton and partly on photosynthates translocated from their algal symbionts; the relative contributions from each source will vary. Corals also produce copious amounts of mucus which is rich in wax esters and triglycerides. Some of this is consumed directly by zooplanktonic copepods and by crabs and fish. Much of it, however, ultimately becomes colonized by bacteria and incorporated into detritus. There can be little doubt therefore that bacteria-laden detritus resulting both from the breakdown of plant material and from coral mucus constitutes a particularly important pathway for the transference of nutrients and energy within the coral-reef ecosystem.

CHAPTER SEVEN

MARSHES AND MANGROVES

In many places where soils and marine sediments meet, a specialized community of terrestrial *halophytic* (salt-loving) plants and associated animals develop (Figure 7.1). This happens mainly in the supralittoral and the upper littoral zones. Two of the most characteristic coastal communities of this nature are saltmarshes, which are found mainly in

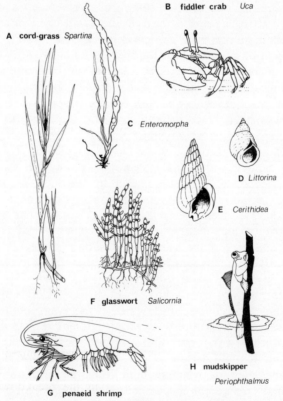

Figure 7.1 Some characteristic species of saltmarsh and mangrove communities (redrawn from various sources). Not to scale.

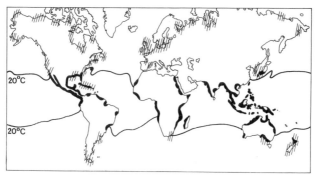

Figure 7.2 The world distribution of saltmarshes (hatched) and mangroves (solid). (After Chapman, 1977.)

temperate to subpolar climates, and mangrove swamps which are of tropical to subtropical occurrence (Figure 7.2).

Saltmarshes

Beeftink (1977) defines saltmarshes as 'natural or semi-natural halophytic grassland and dwarf brushwood on the alluvial sediments bordering saline water bodies whose water level fluctuates either tidally or non-tidally'. In this context the term *alluvial* includes material left by the sea as well as by fresh water. Saltmarshes occur where the average salinity is in the range of 5–38‰. In the higher reaches of estuaries true saltmarsh is replaced by reed and rush beds or other fringing vegetation. Such transition can also be seen on moving directly inshore in many marshes. In a few subtropical areas marsh and mangroves coexist, for example on parts of the southern Australian coast marshes are fronted by white mangroves on their seaward border. Such mixed systems also exist in the Caribbean and on some dry tropical coasts. Although saltmarshes occur on fairly open and on enclosed coastlines, some degree of shelter is necessary. This is because their development entails a net accumulation of sediment. Indeed the water and sediment regimes are the two prime factors in marsh establishment, growth and differentiation (see Ranwell, 1972; Long and Mason, 1983).

Establishment

The dominant organisms of saltmarshes are rooted plants of terrestrial evolutionary origin. Colonization by such plants cannot occur unless the sediment is stable enough to permit rooting and growth. Coarse to medium sands are not conducive to marsh formation, partly because they tend to dry out. There is a requirement for some admixture of silt. Thus the early stages of marsh development occur in high beach-level mud and silty sand. These areas tend to be flat and slow-draining—features which favour the formation of a surface film or mat consisting of protists, prokaryotes and filamentous algae. Coles (in Jefferies and Davies, 1979) demonstrated the importance in these early stages of epipelic diatoms; their secretions trap and bind silt particles together. Sand-flat areas of the Wash, England, had few such algal protists, but removal of macrofauna led to increase of these microalgae and increased accumulation of silt. Conversely, removal of microalgal film from the mud-flat and saltmarsh surface interrupted sediment deposition. Detached seaweed and dead shells on mud-flats can form depositional centres producing small hummocks or banks. Silty sand can be bound together by the secretions of numerous tube-dwelling polychaetes such as the spionid *Pygospio*. However, little is known about the role of macro- and meio-faunal species in high-shore depositional processes. As centres of deposition and accretion come to stand slightly higher than the general sediment the surface drainage is concentrated into the lower areas so that a system of small channels is formed. These gradually become better defined, partly by scour but particularly by depositional processes continuing to add sediment elsewhere.

At some stage during this process the first angiosperms colonize the raised sediment surface. This is either through germination or occasionally through transport of vegetative material. Sediment temperature range will be less in the lower parts of saltmarshes due to amelioration by the sea ; the sediment will be permanently moist and the salinity usually close to that of the adjacent sea. The two genera which are most prominent as pioneer saltmarsh plants are *Salicornia*, often known as glasswort, samphire or pickle-weed, and *Spartina* or cord-grass. The latter is absent, except as an introduction, from some areas, for example New Zealand and western South America. *Puccinellia, Scirpus* and *Juncus* species are also commonly present.

Salicornia species are predominantly annuals. The leaves have become modified to produce stem-like structures (phylloclades) which are probably less prone to wind and wave damage and perhaps aid water balance. Some glasswort species are unable to tolerate low salinity. Dispersal is by individual seeds or by entire seed heads. In some species, such as the Californian *S. bigelowii*, seeds may float for several days or even months, in others a proportion of seeds remains on old plants and germinates after burial of these *in situ*. *Salicornia* colonization often occurs in sediments with high positive surface redox potential, and it is thought that oxygen production by algal protists such as the diatom *Pleurosigma* may aid seedling growth. Glasswort produces a rather sparse vegetational cover especially when first colonizing an area; however, its presence slows current speed and thus aids further sediment deposition. This is also increased where filamentous or ribbon-like algae such as *Bostrychia* and *Enteromorpha* grow among or are tangled with the glasswort. Many seedlings are lost by their inability to stand the tidal or wave-induced currents ebbing and flowing across the sediment; the mortality near the HWN mark may be almost double that at HWS level.

Spartina is a 'prosperous' grass genus, with fifteen or more species, specializing in saltmarsh life. Several of the species are *primary colonizers* in marsh formation. They include *Spartina alterniflora* on the United States' eastern seaboard and *S. anglica* on European shores. The latter species is a fertile hybrid which arose in Southampton Water in the mid- to late nineteenth century from a complex of sterile and fertile hybrids (generally referred to as *S. townsendii*) of the American *S. alterniflora* and the European *S. maritima*. Colonization is mainly by seed dispersal and germination. Seeds of *S. anglica* can grow from beneath 10 cm or less of mud, and as in *Salicornia* shallow-rooted seedlings are often lost through tidal action. *Spartina anglica* seedlings do not survive shaded conditions and this may restrict the plants' distribution amongst the taller landward vegetation. Plants can withstand long periods of submergence but this inhibits flowering. Cold winters can check growth severely. Once established, *Spartina* can spread rapidly by its underground tiller system; the plants also have deep anchor roots. Cord-grass has large internal air spaces and recent work has shown that oxygen is released, both passively and actively through metabolism, into the sediment, thereby increasing Eh around the roots. This in turn favours more rapid plant growth.

Succession, zonation and marsh structure

Various other plants can act as primary colonizers, for example *Vaccinellia* in Europe or *Scirpus* in New Zealand. Following primary colonization other species can become established due to environmental changes induced or accentuated by the pioneers. These changes and the subsequent arrival of further species may lead to the habitat becoming unsuitable for some or all of the earlier species. Thus a series of fairly predictable changes in community composition, termed a *succession*, is induced. In saltmarshes the dominant changes are associated with sediment accretion, leading to raising of the marsh surface and thus to an increase in drainage and to a decreased submersion in salt water. These are also factors which change with distance inland (and hence height) from the marsh edge. The succession pattern is therefore often paralleled by the marsh zonation. Thus in saltmarshes temporal events may mirror spatial patterns. Zonation and actual or inferred succession have been studied in many areas. The usual net result is the description of a three- or four-stage system consisting of pioneers on the upper mud-flat which may be bounded by a low salting cliff, a fairly level general saltmarsh with *creeks* and *pans* (saltmarsh pools) and often with definable lower and upper zones and, finally, a high marsh zone bordering more terrestrial shrubby vegetation (Figure 7.3).

The shrub vegetation may contain some characteristic species, such as the Far Eastern *Paliurus ramosissimus* with its sea-distributed fruits (Nakanishi, 1981), as well as species found in other habitats. In tidal areas the boundary between the lower and upper general marsh zone appears to be related to the frequency of submersion. Chapman (1960) therefore termed these zones 'submergence and emergence marshes' and defined the border as corresponding approximately to the level at which there were 360 submergences per annum and a maximum continuous emergence of 9 days. It should be noted that, because of the length of exposure and thus of evaporative periods, salinity may at times be higher in the upper than in the lower marsh. The terms *slobland* and *salting* are used respectively for the upper tidal mud-flat with sea-grasses and pioneer halophytes and for the general saltmarsh with its creeks and pans. In general, high-latitude

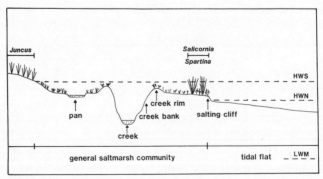

Figure 7.3 Generalized transect of a North European saltmarsh. (After Ranwell, 1972, and other sources.)

saltmarshes have a denser covering of plants than do lower-latitude marshes.

As Ranwell (1967) points out, the chances of any one particular part of a saltmarsh passing through an orderly textbook succession with time may be very low. By the nature of their development marshes are intersected with creeks. These and other features provide a variety of sub-habitats with their own array of wildlife. Sediment trapping by vegetation on the rim of drainage channels may lead to the surface between channels becoming concave but the steepened banks may erode. Tidal currents tend to scour the bank material and whole sections can be undermined and slump into the creek bed. In some marshes it has been estimated that fiddler crabs excavate as much as 20% of the bank face area; such biological activity also contributes to erosion (Letzsch and Frey, 1980). Conversely, between-channel islands may coalesce or channels become blocked to produce primary or secondary pans respectively.

Other flora and fauna

Prokaryotes and protists are important within the general saltmarsh. Several recent studies have examined their distribution and transport (Rublee et al., 1983). There is a high bacterial population density (up to 10^9 cm^{-3}) in sediment and in its covering water (10^6 cm^{-3}). Algal mats, dominated by cyanophytes in the summer and by green algae in spring and autumn, are of common occurrence (Zedler, 1982). Some red and brown algae have adapted to saltings, for example *Bostrychia* attaches to other saltmarsh plants, and some genera including *Fucus* and *Pelvetia* have species with dwarf forms occurring at quite high marsh levels. It is known that saltmarsh soils are rich in fungal hyphae, some of which are associated with halophyte roots.

Various distributional, physiological and manipulative studies of marsh micro- and meiofauna have been made, but knowledge of these groups is still fragmentary. Work in the south-eastern United States shows that the meiofaunal communities are complex (e.g. Bell et al., 1978). Many of the nematodes have an anaerobic capacity suitable for life in sediments having little or no free oxygen. There are few, if any, accounts of individual marshes embracing all components of the wildlife, especially if fish, birds, and mammals are included.

Teal (1962) presents an overall study of marsh energy flow and lists seven main categories of invertebrate fauna, namely *terrestrial* species from the general marsh (1) or from its upper limits (2), *aquatic* species from the seaward edge (3), from creek sides (4) and from the general marsh (5) and marsh-evolved species, either with (6) or without (7) a planktonic stage. The terrestrial components often include one or more large herbivores. Various geese and duck species, rats, moles, rabbits and hares also graze on saltmarshes and can have a controlling effect on the vegetation. It has been shown in some North Carolina marshes that geese can remove over half the plant biomass (Smith, 1982). Smaller terrestrial herbivores such as aphids, grasshoppers, thrips and moth larvae, for example in Europe the shore wainscot *Leuconia littoralis* feeding on *Juncus maritimus*, are also important. One grasshopper species, *Orchelimum fidicinium*, together with a leaf-hopper, is responsible for grazing nearly 5% of *Spartina* production in a marsh in Georgia, USA. The higher marsh often contains one or more herbivorous pulmonate snail species.

Terrestrial predators of the general saltmarsh include hunting wolf-spiders as well as web-spinning forms. Jumping bugs (Saldidae) may also be common. Long-legged dolichopodid flies often replace the seaweed flies of rocky and sandy shores as the main predatory insects. Most marsh dolichopodids feed on other insects trapped at the water–air interface, but at least one species, *Hydrophorus oceanus*, fishes into the water for prey. Various lepidopterans, flies and midges are native to saltmarshes, and unfortunately some areas are also used by fever-spreading mosquitoes. Perhaps these, like, birds should be included in an eighth category—*aerial* marsh animals. Herons, avocets, spoonbills and many other species feed in marsh pools. Waders often use saltings as roost areas during high-tide periods. Some terrestrial invertebrates escape flooding by climbing vegetation but others are able to survive temporary submersion. Some species use air which has been trapped by the soil; a few ant species trap bubbles. A few terrestrially-derived species, for instance of enchytraeid worm, beetle and springtail, apparently live in the permanently wet soil of the lower marsh, but often forms such as earwigs, centipedes and millipedes are only found in the highest marsh levels as part of a fauna similar to that of adjacent soil.

The aquatic fauna is predominantly marine with close affinities to mud- and sand-flat communities. However, its environment is less stable due to temperature, salinity and other changes. Species able to migrate in and out with the tide or able to burrow in the channel banks will experience relatively little change. Creek channels are an important fish habitat, for example gobies, menhaden, mullet and mummichog occur here, though not necessarily throughout their life. Several shrimp (or prawn) species are characteristic channel inhabitants. Tidally-migrating species include many copepods, amphipods and isopods. The small marsh goby *Pomatoschistus microps* is sometimes infested with the externally parasitic larvae (praniza) of the isopod *Paragnathia formica*. The adults, which are non-feeding, live in small burrows dug by the males high in the channel banks or salting cliff. The females die after shedding the larvae, which then leave the burrows to infest the fish. Other small high-level burrows may belong to amphipods, beetles or other terrestrial or semi-terrestrial species. Small low-level burrows are inhabited by marine amphipods, for example *Corophium*, isopods and a variety of euryhaline polychaetes. Some animals, including various small mud-snails such as *Hydrobia*, enter the substratum temporarily, presumably to avoid desiccation or to obtain food.

More obvious channel burrows often belong to crabs. The only European species found in marshes (particularly when young) is the common shore crab *Carcinus maenas*, but in lower latitudes many crabs have specialized to marshes and other estuarine or inland habitats. Among the best known are sesarmids and fiddler crabs of the genus *Uca*. The latter have been the subject of many comparative distributional, physiological and behavioural studies. Fiddler crabs often scurry along creek banks in fairly large herds. The crabs orient to visual stimuli including the sun's position and the dominant plane of light polarization (water-reflected light tends to polarize). Fiddlers use sound in their courtship behaviour and also communicate visually by species-characteristic waving of their enlarged cheliped. Conditions in fiddler-crab burrows can become micro-oxic or anaerobic, but the crabs' respiration rate is independent of oxygen levels until very low concentrations are reached and there is also a capacity for anaerobic respiration involving glycolysis. Detritus and sediment bacteria are probably the main food source but it is thought that *Uca pugnax* may feed selectively on meiofauna. It has been shown that nematode density is greater around the burrows than elsewhere, whereas copepod abundance is less, but the reasons for this are not established.

A few species of nudibranchs are known from saltmarshes. It is less surprising that various shelled molluscs are able to survive in marshes, since they are able to shut themselves away from temporarily adverse conditions. Although they are found mostly in creeks, some species extend their distribution into the pans. In Europe these tend to be species that are also found on mud-flats, for example *Mya arenaria* and *Mytilus edulis*; the same is true of prosobranchs such as *Littorina*. Elsewhere, less eurytopic species such as the North American ribbed mussel and marsh winkle may occur. The latter, *L. irrorata*, mainly grazes on dead *Spartina alterniflora* but also browses on algal mats and sediment. The ribbed mussel *Geukensia* (= *Modiolus*) *demissa* is a filter feeder often found attached to the lower stem of cord-grass or reeds but sometimes lying nearly buried in sediment. In addition to its normal respiration from sea water it can also make use of oxygen from air taken into the mantle cavity.

The fauna of saltmarsh pans is an impoverished version of that of the channels and mud-flats. Pan temperatures and particularly salinities (because of precipitation and evaporation) tend to fluctuate widely; in addition, pans may lose all their surface water so that very few species are able to survive.

Trophic structure and energy flow

It is possible to distinguish four main lines of current research into saltmarsh ecology. These are (1) continuing description and classification of the major components of previously undescribed marshes; (2) the detailing of subcomponents, particularly microflora and meiofauna and their function; (3) studies of the effects of managing or altering controlling factors such as predation, grazing, nutrient levels and flooding; and (4) budgeting of energy and material flow through the community. Perhaps the latter subject has aroused the most widespread interest.

Most of the work on saltmarsh processes has been conducted in the eastern United States. This is hardly surprising, since in South Carolina and Georgia for example some 20–30% of the shore is marsh. By the late 1960s three originally reasonable hypotheses were in danger of becoming dogma. They were (1) *Spartina* marshes were the most productive marine ecosystem; (2) detritus was the dominant food source within the marsh; and (3) the majority of the detritus was exported and 'fuelled' the adjoining coastal waters. Recent work (see Nixon in Hamilton and Macdonald, 1980) shows that more caution should have been taken in extrapolating the early results to all coastal marshes; as one might expect from a natural system, there is considerable variability in the generation and fate of organic products. American east-coast marsh production can attain over $500 \, \mathrm{g\,C\,m^{-2}\,y^{-1}}$, as can some European marshes. The average primary production of saltmarshes is probably between $200–400 \, \mathrm{g\,C\,m^{-2}\,y^{-1}}$. In some marshes *Spartina* may contribute 50% of this, but sedimentary and planktonic algae are also of importance. The overall primary productivity is equivalent to about 1.5% of the incident light energy and is probably equalled by some kelp beds and upwelling systems. Nitrogen is often a limiting factor and the rate of import of nutrients, for example, in sediment brought by tidal flooding, is important. However, tidal exchange can result in a net loss of nitrogen from marshes. This is balanced by nitrogen (chiefly as nitrate) input by fresh water and by some *in-situ* bacterial fixation. *Spartina* has nitrogen-fixing bacteria on its roots and stems and also in its cortical layer.

There are two major pathways of energy transfer from primary producers to heterotrophs in saltmarshes. These are by grazing vegetation directly and by consuming (or grazing) detritus. Although it is clear that the detritus route is dominant in some marshes, with direct herbivory consuming 10% or less of the primary production (but *vide* birds and mammals), it is probable that detritivores derive their food mainly from bacteria and other microfloral and meiofaunal components associated with the plant material. *Spartina* is at least as refractory as *Thalassia* (see Chapter 4). Saltmarsh plants also lose some organic molecules by leaching.

It is true that some marshes export considerable quantities of plant debris to coastal waters. In some northern saltmarshes most of the above-ground vegetation is carried to sea by ice. However, as is obvious in the latter case, the export process is often seasonal and may be replaced by import at other times of year. In some marshes over 75% of the plant material may be retained in the marsh by heterotrophic and sedimentary processes. Some recent studies have made use of the relative abundances of naturally-occurring carbon isotopes to help follow the fate of decaying *Spartina*. Although the value of saltmarshes in exporting productivity may have been exaggerated, they also function as nursery areas for many fish and crustaceans. Thus there is no doubt that saltmarshes play an important and intrinsic role in the biology of coastal areas. A saltmarsh food-web and energy-flow diagram is presented in Figure 7.4.

Mangroves

Mangroves form dense thickets consisting of some of the few terrestrial plants capable of tolerating fully saline conditions. They occur along protected sedimentary shores especially in tidal lagoons, embayments and estuaries (see McNae, 1968). Although they can penetrate far inland they are never totally isolated from the sea. The rigours of life at the interface between land and sea has produced a vegetation which is of relatively low diversity when compared with other tropical and indeed temperate plant communities. Their emergent, evergreen canopies are inhabited by a

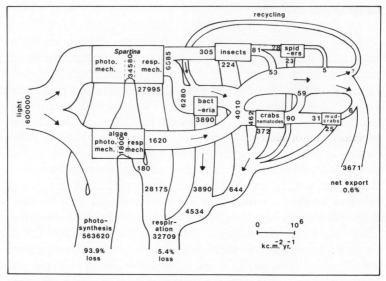

Figure 7.4   Energy flow through a saltmarsh, Georgia, USA. (After Teal, 1962.)

variety of terrestrial animals including insects, birds and arboreal mammals, whilst their complex root systems, and the attendant thick silty sediments in which they grow, provide suitable habitats for many marine invertebrates. In addition, numerous fish and crustaceans move in and out of the mangrove system with the tides. Mats of filamentous algae (e.g. *Cladophora, Vaucheria*) present on the substratum trap fine sediments, and leafy algae (*Bostrychia, Gelidium*) attach directly to the mangrove roots.

The term mangrove has been used to describe both an ecological group of flowering halophytic shrubs and trees (up to 30 m high) belonging to several unrelated families, and the complete community or association of plants which fringe sheltered tropical shores. The term 'mangal' is sometimes used to describe the latter. It is estimated that some 60–75% of tropical coastlines may be fringed with mangroves (Walsh in Reimold and Queen, 1974). They are important to man as a source of timber, charcoal and tannin, their stabilizing properties protect coastlines from major erosional damage by tropical storms and they provide an important nursery ground for many commercially important fish and shellfish. For an account of the economic value of mangroves see Christensen (1983).

Geographical distribution

Although mangroves attain their maximum development and diversity along moist warm coastlines of the tropics where rainforests are the climax vegetation, some species (e.g. *Avicennia marina*) do extend northwards to Japan and Bermuda and southwards to Australia and northern New Zealand (Figure 7.2). Their geographical limits appear to be set by low temperature. Sensitivity to temperature varies with species, but mangroves generally cannot survive air temperatures below $-4°C$. Optimal climatic conditions occur where the temperature rarely falls below $20°C$ and where seasonal variations are $< 5°C$. Dispersal ability is also clearly an important factor determining geographical distribution. The capacity for some seedlings to disperse over long distances may account for the wide distribution of certain mangrove species (e.g. *Rhizophora mangle*).

Two main groups of mangroves are recognized, those from the Indo-Pacific region and those from the Atlantic. Only a few genera (no species) are present in both regions. Considerably more species and genera occur in the Indo-Pacific (60 species compared with 10 species in the Atlantic province). The greater diversity of species in the Indo-Pacific (compare corals, Chapter 6) has led to the suggestion that mangroves may have spread outwards from that geographical region. This 'Centre of Origin Theory' however, is largely inconsistent with the fossil record (McCoy and Heck, 1976). Alternatively, the greater diversity could reflect the general lack of long-distance dispersal (see p. 124) by the propagules of most mangrove species, together with the greater opportunity for speciation along the extensive coastline provided by the numerous geographically isolated islands in that biogeographical province.

Mangroves may be locally absent within the latitudinal range of the community type for various reasons. The forces responsible for such regional distributional patterns fall into two broad categories: those which *prevent* propagules from becoming established (e.g. unfavourable soil conditions; insufficient protection against tidal currents; adverse shore profile) and those which *disrupt* established mangroves (e.g. fires; hurricanes; coastal erosion). Within any given geographical region local conditions will therefore favour some species but not others.

Zonation

One of the most distinctive features of mangrove vegetation is the occurrence of species or groups of species in discrete bands or zones

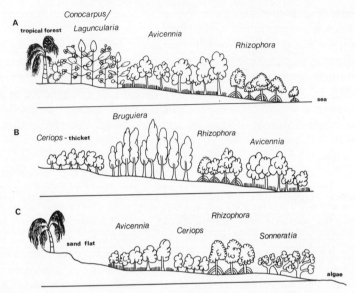

Figure 7.5 Generalized zonation patterns of mangroves in different geographical regions: (*A*) Florida; (*B*) West Indo-Pacific; (*C*) East Africa.

running across the marine–terrestrial environmental gradient. Each zone is typically dominated by a single species. Although considerable emphasis has been focused on detailed descriptions of the species composition and arrangement of these zones in the major mangrove systems of the world, no universal zonation scheme such as that described for rocky shores (p. 40) has so far emerged. Species composition varies and even whole zones may be transposed between one area and another (Figure 7.5). In areas where mangroves are restricted in area (e.g. where the tidal range is small) no zonation patterns may be discernible. At the limits of mangrove distribution, few species may be represented, with consequential implications for zonation. Numerous physical and biological factors have been regarded as important determinants of mangrove zonation. These include the following.

(1) Tidal regime

This is possibly the single most important factor in determining mangrove zonation. The number of consecutive days that a particular zone is exposed by the tides seems to be especially important for the establishment and subsequent survival of mangrove propagules. Aerial exposure dries the soil and leads to an increase in salt content. As very high salt concentrations develop in the upper soil layers, mangroves tend to be replaced progressively by halophytes like *Salicornia*, and in extreme cases a bare salt desert on the landward margins of the mangrove forest may develop. Tidal flushing also influences production, in particular by transporting oxygen and nutrients to the root zone and by removing toxic sulphides and excessive amounts of salt. Accretion and erosion are also related to tides.

(2) Salinity

Evidence suggests that mangroves are facultative rather than obligate halophytes. Their tolerance of saline conditions, however, enables them to survive in environments which are unavailable to their principal competitors (e.g. tropical rain forests). It has been suggested, therefore, that salinity may simply serve as a 'competition eliminator' (Thom, 1967) and not a principal determinant of zonation. Salt concentration in the substratum is highly variable and depends on evaporation, tidal flushing and freshwater inputs from rainfall, subterranean seepage and terrestrial run-off. Species vary in their tolerance or adaptation to salinity (Bunt *et al.*, 1982) and each accordingly occupies the appropriate zone on the environmental gradient. Even within any given species, seedlings and established plants may differ in their relationship to salinity. Soil salinity influences growth (and therefore size) through its effect on photosynthesis and respiration rates. Hypersaline conditions lead to stunting. Salt spray may be important at the seaward margins of the mangrove swamp.

(3) Geomorphology and climate

These are important largely through their effects on other factors (see, e.g., Semenuik, 1983). The nature of the sediment (e.g. degree of aeration, water content), the pattern of tidal inundation, drainage and soil salinity all depend to a large extent on local topography. Erosion can truncate or even alter zonation patterns. Mangroves are often most luxuriant in areas of high precipitation. Rainfall is especially important at the landward margins where aridity and high salt content of the soil during neap tides can potentially exclude mangroves. High temperature increases evaporation which in turn increases soil salinity; it also accelerates the rate of oxygen consumption by micro-organisms and can thus lead to anoxic conditions.

(4) Species interactions

Studies on mangrove zonation have concentrated on the importance of the physical environment and much less is known about the biological determinants of zonation. Competitive ability is probably important in determining which species become established within any particular zone. Dense canopies of *Rhizophora* and *Ceriops* considerably reduce the amount of light reaching the understorey and prevent colonization of open-canopied forms like *Avicennia* whose sporelings are less tolerant of shade. Sometimes species share the same optimal environment and the outcome then depends on relative competitive ability. *Rhizophora* and *Laguncularia*, for example, achieve maximum growth in the intertidal zone in Florida (Ball, 1980). When these occur concurrently *Laguncularia* is eventually forced into suboptimal areas above high-water mark. Such competitive exclusion of one species by another may be widespread, though detailed investigations are generally wanting. The fauna may also be important in determining species distribution patterns. Beetle larvae (e.g. *Chrysobothus*) feed on and kill *Rhizophora mangle* and the subsequent build-up of plant debris makes the environment more suitable for *Avicennia germinans*. Some crabs strip bark and consume mangrove seedlings. The isopod *Sphaeroma terebrans* bores into prop roots; damaged trees may be severely weakened and thus more easily eroded by waves and currents.

(5) Man's influence

Stresses induced by man are usually more intense and non-selective than those which occur naturally. Mangroves are cut for their tannin properties, for construction timber and for charcoal. Excessive logging can lead to changes in species composition. Mangrove swamps act as sediment traps— when destroyed this sediment may be carried out to sea where it can smother neighbouring coral reefs. Drainage schemes, thermal stress from power plants and the extensive use of defoliants (e.g. during the Vietnam War) all influence the local distribution patterns of mangroves.

(6) Stochastic (= chance) factors

Sudden unpredictable catastrophes can befall any community and bring about gross changes in species composition. Hurricanes and lightning fires, apart from their direct destructive effects, open up the mangrove canopy and thus enable light-demanding species to become established. They may therefore serve to maintain diversity in much the same way as physical disturbance and predation prevent competitive monopolization on rocky shores (p. 49). However, in Florida, areas of mangrove destroyed by hurricanes were subsequently invaded by dense stands of the fern *Acrostichum aureum* which acts as a barrier to recolonization by more light-demanding mangrove seedlings. The effects of hurricanes are variable and depend on wind direction and velocity, local topography and vegetation types; some species (e.g. *Avicennia germinans*) are especially resistant and it is easy to envisage that under a regime of regular hurricanes it could become a dominant species. There may be a degree of randomness in mangrove colonization, the species composition of seedlings in virgin areas of Florida everglades, for example, simply reflecting that of the propagules that 'happened to arrive' (Ball, 1980).

Several factors are therefore probably involved in determining mangrove zonation patterns. It seems unlikely that greater emphasis can be attached to any one particular factor since none of the physical factors is totally independent of the others. Changes in one factor lead to changes in other factors with inevitable consequences to the flora. The degree of importance attached to any one factor therefore probably varies from one area to another, depending on local conditions and species occurrence. However, one factor common to all mangrove communities is tidal inundation. The frequency and duration of inundation is often well correlated with zonation. Tides exert an over-riding influence through a multitude of intermediate factors, and these have a *collective* influence with a gradient roughly normal to the shoreline. Each species will therefore find its own particular place within this environmental gradient. The general pattern may be complicated by creeks, hollows or hummocks which produce *local* changes in environmental conditions and therefore corresponding changes in vegetation. Physical factors must also ultimately determine the relative competitive abilities of component species and it is probably a combination of physical and biological factors that determine the distinctive zonation patterns of mangroves. A multivariate approach to this problem is clearly long overdue.

Succession

One of the most controversial areas of mangrove ecology is whether zonation of species across the marine–terrestrial gradient constitutes true *succession.* Some authorities argue that zones do indeed represent serial stages in the succession from an estuarine/mud-flat type of community to a climatic climax community (in this case a tropical rain forest) though any one of the later stages in the succession may be repeated before the succession eventually proceeds to the next stage. In some cases, physiographic conditions may not be suitable for transition to the tropical forest and a preclimax swamp forest effectively becomes the climax community. In this interpretation of zonation, the coastal zone is the pioneer stage and the more landward zones progressively later stages in the succession. The concept implies a gradual seaward movement of the system (hence the reputation of mangroves as land builders) and that mangrove vegetation has the ability to modify the habitat (e.g. by accumulating sediment) thus inducing subsequent vegetational changes. One community, therefore, succeeds another until an emergent forest is formed.

This classical view of zonation has been criticized largely on the grounds that it is inconsistent with geological data. An alternative explanation is that zonation represents a response of the mangrove ecosystem to external forces rather than a temporal sequence induced by the plants themselves. Mangroves only grow where sedimentary regimes are suitable, they therefore follow sedimentation rather than cause it, at least initially. However, once established, their net of prop roots acts as a sedimentary weir, slowing water movement and allowing any load of entrained particles to settle out. As with most controversial issues the truth probably lies somewhere between the two views. Certainly mangrove vegetation is sensitive to changes in the physical environment. Classical serial succession, however, may provide an appropriate generalization for many mangrove communities.

Adaptation of mangroves

Mangroves exhibit numerous physiological and structural adaptations to their unusual physical environment. Consequently they provide one of the most remarkable examples within the plant kingdom of convergent evolution between several taxonomically unrelated groups (12 or so genera from 8 families) living within the same habitat. Mangroves become established on sedimentary shores where wave action is minimal. The lack of vigorous water movement results in the gradual accumulation of fine-grained muds, a process which is subsequently enhanced by the mangroves themselves. Mangroves must, therefore, be adapted not only to salinity stress but also to growth in waterlogged, anoxic sediments.

All mangroves have evolved a system of shallow, laterally-spreading cable roots. This underground system is richly supplied with finer anchor and absorptive roots. Above-ground roots are of two main types—pneumatophores (e.g. *Avicennia nitida*) and prop roots (e.g. *Rhizophora mangle*). Pneumatophores are negatively geotropic and extend upwards through the soil surface (Figure 7.6). They arise from the cable root system and may be branched or unbranched. The extensive development of pneumatophores enables species like *A. nitida* to grow in extremely anoxic sediments. Stilt or prop roots (= rhizophores) emerge from the trunk well above ground and curve downwards into the surface mud. They provide firm anchorage and allow exploitation of particularly soft sediments. The type of root system will therefore to some extent serve to determine zonational position. Both types of roots contain numerous small pores or *lenticels* which are permeable to air but not to water. These communicate with an elaborate system of *aerenchyma* tissue and thereby provide rapid transport of gases to the underground root system. Oxygen within these intercellular passages is utilized when the roots are submerged. If the lenticels are experimentally coated with grease then the oxygen within the root system is rapidly depleted.

Halophytes in general are faced with the major problem of absorbing

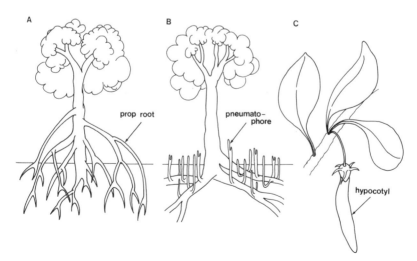

Figure 7.6 (*A*) Prop roots of the red mangrove (*Rhizophora*); (*B*) pneumatophores of the black mangrove (*Avicennia*); (*C*) mangrove propagule.

E

water against the osmotic pressure gradient that exists between the saline sediment water and the plant vascular system. Their survival therefore depends on their ability either to accumulate high concentrations of salt (e.g. *Avicennia marina*) which they subsequently eliminate through special salt-secreting glands in the leaves, or to exclude salt from the roots by a process of ultrafiltration at the cell membrane (e.g. *Bruguiera gymnorrhiza*). In the latter case any salt that does enter the plant can be stored in old, succulent leaves which are periodically shed. The complex physiological mechanisms underlying these responses are reviewed by Scholander (1968).

The osmotic pressure gradient between soil water and plant sap controls the rate at which water is supplied to the leaf tissue in order to support transpirational water loss. When this supply is deficient (e.g. at higher salt concentrations), the stomata, which are set in pits on thick waxy leaves (xeromorphic adaptations) close in order to conserve water. This, however, also reduces the uptake of gaseous carbon dioxide, which in turn depresses photosynthesis and presumably therefore primary production. Succulence, associated with the presence of water storage tissue in the leaves, is a further adaptation to the *physiological dryness* of the mangrove habitat.

Mangroves therefore have several ways of controlling salt concentration and it is perhaps not surprising that different species will vary in their salinity tolerances, with consequential implications for local distribution patterns.

Many mangroves (e.g. *Rhizophora, Bruguiera, Ceriops*) are viviparous, having unusual 'seeds' that actually germinate whilst still attached to the tree (Rabinowitz, 1978). Each propagule (Figure 7.6) produces a long torpedo-shaped fruit (= hypocotyl) and when these are released they plummet vertically into the mud and immediately become established. In some species, however, they float away on the tide, sometimes being transported over long distances by ocean currents. The dispersal properties of propagules also appear to be correlated with mangrove zonation. Some mangroves (e.g. *Avicennia*) are not viviparous but have seedlings that can withstand prolonged periods of tidal inundation by virtue of their ability to respire anaerobically.

Mangrove fauna

The fauna of mangroves, derived largely from adjacent terrestrial and marine habitats, has been less well studied than the mangrove vegetation itself. Broad patterns of zonation can be discerned both horizontally through the swamp and vertically from the sediment to the emergent canopy. Vertical stratification depends mainly on tidal inundation and salinity. The canopy is largely free from tidal influences and supports a fauna (insects, birds, mammals) that is essentially of terrestrial origin. These species generally show no special adaptations for life in mangroves, though many of them do feed on the marine species below. They also contribute to the nutrient input into the mangrove ecosystem in the form of faecal material. Root-holes and clefts retain rainwater and provide a valuable microhabitat, especially for insects. Below the canopy salt-tolerant species appear. The distribution of these essentially marine species depends primarily on their resistance to water loss and the availability of food and suitable substrata. Most species exhibit distinct preferences for particular habitat types which also occur elsewhere, and their association with mangroves is therefore often considered to be rather fortuitous. McNae (1968) provides an interesting general account of the macrofauna of Indo-Pacific mangroves.

The marine component of the mangrove fauna is largely dominated by various molluscs, crustaceans and fish. Littorinid snails (e.g. *Littorina scabra*) and many sedentary invertebrates (e.g. oysters, ascidians, sponges, barnacles) are especially prominent on the trunks and extensive root systems. The mangrove snail *Cerithidea* feeds on the mud surface at low tide but climbs up the mangrove trunks with the rising tide to escape predators, mainly crabs and fish. The snails appear to have an inbuilt biological rhythm, since their upward migration starts before the tide reaches them. Lucinid lamellibranchs may be locally abundant in mangrove sediments. Prawns (e.g. *Upogebia*), fiddler crabs (e.g. *Uca*), tropical land crabs (e.g. *Cardisoma*) and ghost crabs (e.g. *Dotilla*) burrow into the soft sediments; their burrows enable oxygen to penetrate the sediment and thus ameliorate the generally anoxic conditions of mangrove muds. Many of these crabs have mouthparts specialized for straining particles of organic detritus from the mud. Some show partial adaptation for air-breathing, the walls of their gill chambers being highly vascularized and lung-like. *Sesarme eulime* and *S. meinerti* are desiccation-resistant species and occupy the drier areas of the landward margins; *S. guttata* and *S. catenata* are confined to shaded areas due to their inability to resist water loss. Mudskippers (e.g. *Periophthalmus*), a group of unusual gobioid fish, climb over roots and walk on their modified pectoral fins, or 'skip' using their fins and tail. Their large, frog-like eyes, set on top of the head, function best in air, whilst vascularized sacs present in the mouth and gill chambers enable them to breathe air. Mullet (*Mugil*), swimming crabs (e.g. *Callinectes*) and penaeid prawns visit the mangrove at high tide. Snakes, lizards and crocodiles may also be present (Chapter 9).

Primary productivity; trophic interrelationships

Despite the dearth of quantitative studies, available evidence strongly suggests that mangroves are highly productive systems. Estimates indicate that net primary production may be around $350–500\,g\,C\,m^{-2}\,y^{-1}$; phytoplankton production in adjacent nutrient-deficient tropical waters, on the other hand, rarely exceeds $50–75\,g\,C\,m^{-2}\,y^{-1}$.

Much of this primary production eventually enters the aquatic system as plant debris (e.g. leaves, twigs, scales). This is attacked by micro-organisms and is subsequently consumed by a wide variety of detritivores which fall broadly into three functional groups: (1) grinders (= shredders) which chew large leaf particles; (2) deposit feeders, which select smaller particles deposited on the sediment surface; and (3) filter feeders, which strain fine particles from the water. The detrital food chain of the Florida red mangrove (*Rhizophora*) has been extensively studied (e.g. Odum and Heald, 1975). The cyclical pattern of this chain (Figure 7.7) arises from the multiple re-use of detritus. Detritivores digest the micro-organisms present on the surfaces of the decaying plant material before returning finely macerated debris to the environment as faecal pellets. This is subsequently reworked by micro-organisms (= 'faecal loop') leading to further breakdown, and the whole cycle is repeated. The final stages in the main food web are small carnivores (e.g. minnows) and higher carnivores (fish-eating birds and 'game' fish). Black mangroves (*Avicennia*) occur at higher tidal levels in the Florida mangrove swamp than does *Rhizophora*. Mosquito and midge larvae are the principal detritivores in this zone (Figure 7.7) and these in turn are consumed by a variety of 'forage' fish.

Mangroves are interface ecosystems with no precise boundaries to distinguish them from neighbouring systems on which they depend (see Lugo and Snedaker, 1974). Nutrient cycles are driven by various physical and biological factors that control the rate of import and export of inorganic and organic materials. Of these, tidal flushing, riverine input, accumulation and decomposition of plant detritus and certain activities of the fauna are probably the most important. Little is known about secondary production in mangroves. However, it does appear that substantial quantities of nutrient may be exported from the system to be consumed

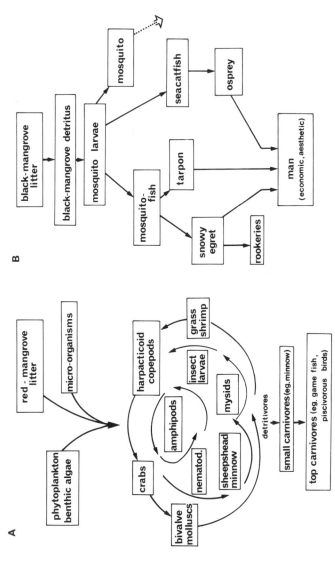

Figure 7.7 Detrital food chains of (*A*) the Florida red mangrove (after Odum and Heald, 1975); (*B*) the black mangrove (after Clark, 1977).

elsewhere; the extent to which this occurs probably varies with local topographical conditions.

Thus although mangroves occur only in a relatively narrow coastal strip (comparable in this respect to the macroalgae which often festoon sheltered rocky shores in cooler latitudes) they are extensively distributed throughout the tropics and their contribution to coastal productivity in waters which otherwise tend to be rather unproductive may be extremely important. Their high levels of primary production enables mangroves to support diverse faunal assemblages which include species of birds and game fish which are of direct commercial and aesthetic importance to man.

CHAPTER EIGHT

OTHER COASTAL HABITATS

In the earlier chapters many of the main habitats of the coastal zone have been considered, but some areas and some organisms have had little or no mention. Habitats such as shingle beaches are only of relatively small and local importance, whereas polar ice-edges or coastline are extensive. The animals so far neglected include fish, birds and mammals including man. This chapter considers some of our 'remaindered' habitats.

Sand dunes

The strandline

Coastal sand dunes, like saltmarshes, result from the stabilization of transported sediment by vegetation. They are formed from wind-blown sand rather than from water-moved silt. Although the development of dunes is wind-dependent, the earliest stage depends on tidal or climatically-induced water-level changes which enable establishment of a high level *drift-* or *strandline*. Sand blown up the shore settles around this drift which contains seaweed or sea-grass debris and often some animal material. Drift lying on the shore promotes water retention in the underlying sediment and provides a supply of nutrients. In addition, temperature fluctuation beneath drift may be dampened to only a third of its original range. There are therefore good germination conditions for salt-tolerant seeds. In northern boreal climates strandline plants tend to germinate in late April and May after deposition of drift by the equinoctial spring tides.

One of the commonest strandline plant genera is *Atriplex*, like *Salicornia* a member of the Chenopodiaceae, a family specializing in coastal habitats. *Atriplex* species, some of which are common to North American and European Atlantic shores, reflect latitudinal and other climatic change; *A. arenaria* extends from Florida to New Hampshire but is replaced to the north by *A. glabriuscula* and *A. sabulosa*. In Europe the latter two species hardly penetrate the Baltic where *A. littoralis* is fairly common in the south and *A. longipes* in the north. *Salsola, Honkenya* and *Cakile* are also widely distributed. Strandline plants are said to have a high nitrogen requirement which is met by bacterial decomposition of the drift, although laboratory work with *Atriplex hastata* has demonstrated survival with very low nitrogen levels. The presence of driftline plants leads to further accumulation of sand, thus raising the beach level so that dune pioneer species can become established. In Europe *Agropyron (Elytrigia) junceiforme* and *Elymus arenarium* are important species. *E. mollis* plays a similar role on some Canadian shores and *Uniola panaliculata* on the United States eastern seaboard. These grasses rapidly produce a vertical and lateral root and rhizome system which binds the sand together. Reproduction is by seed or from rhizome fragments. It is known that *Elymus* seeds have a water-soluble germination inhibitor so that growth is not initiated in dry periods. On germination a vertical seedling root is produced rapidly. This penetrates to deeper and moister sand layers. Lateral roots and then rhizomes are established which spreads the plant outwards.

Accumulations of drifted seaweed provide habitats for a variety of invertebrates. The marine component is limited but usually includes one or more amphipod species; 'terrestrial' species include seaweed-fly larvae, such as *Coelopa*, small oligochaetes, beetles and mites. The species composition varies with size and depth and, at least in Europe, there is a decline in diversity with increasing latitude (Remmert, 1964).

Dune ridge formation

The process of sand transport by wind was studied first in deserts but the same principles apply to beach dune sand. If surface wind is in excess of about $4.5\,m\,s^{-1}$ some sand particles will be lifted from the sand surface. They then fall back, setting other particles into motion so that these appear to jump from the sand surface, a process known as *saltation*. Sand leaving the surface may enter a higher (and more laminar) air stream and thus be carried greater distances before falling. However, the presence of vegetation decreases air flow considerably. As a net result, sand coming into the vicinity of pioneer growth is likely to settle and remain on the sediment surface. Embryonic dunes are formed providing that upward and lateral grass growth can keep pace with deposition (Figure 8.1); upward rates of $30\,cm\,y^{-1}$ have been reported.

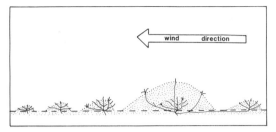

Figure 8.1   The formation of embryonic sand dunes by deposition around an upward growth of *Agropyron*. (After Ranwell, 1972, from Nicholson.)

The pioneer grasses are unable to build high dunes probably because of moisture limitation. Formation of higher dunes is through the action of other grass species, in particular *Ammophila arenaria* and its congener *A. breviligulata* (both commonly known as marram grass). Various other species may be involved in dune building, for example in Japan a sedge and *Wedelia prostrata* (Compositae) are important. *Ammophila*, like the foredune grasses, can establish itself from seeds or from rhizome fragments. The plant is able to maintain an above-ground portion by leaf growth, providing accretion is not rapid. However, with increased deposition *Ammophila* keeps pace by producing long vertical stem (rhizome) internodes which root adventitiously near the sand surface. Horizontal rhizomes are also produced as the plants grow. Previous years' sand accretion rates can be measured from the buried vertical internode lengths since the nodes form at the level of each year's leaf production. Rates of up to $1 \, \mathrm{m \, y^{-1}}$ have been found.

As mentioned, the growth of vegetation slows air flow so that surface wind speed may be reduced below the level necessary for erosion. However, as dunes grow upward they interfere with laminar wind flow. As air is forced upward by the dune face the wind speed increases to reach a maximum near the dune crest. After passing the crest, wind speed lessens and there is usually a lee or backface vortex. Sand carried off the forward slope will be deposited on the lee face. Thus the dune is moved slowly backwards. The rate of movement depends on wind speed, particle size and the degree of protection afforded by vegetation. Values of $9 \, \mathrm{m \, y^{-1}}$ have been recorded in France, but more usual rates are $1 - 5 \, \mathrm{m \, y^{-1}}$. This type of dune movement tends to occur on exposed coasts and where the sand supply is limited; thus it is more characteristic of eroding than of accreting coastline. In an idealized dune system one might expect to find a series of small *foredunes* followed by a series of *dune ridges* and intervening hollows (*slacks*) parallel to the shoreline. If the direction of the strongest and of the prevailing winds coincide, the highest dunes lie some distance inland. If these winds are opposed with the prevailing wind being offshore the highest dunes will lie near the beach. Due to the uneven cover of vegetation, dune crests rarely erode evenly and as the wind is funnelled into eroding areas an extensive 'blow-out' may be produced. This may have the effect of bending part of the main dune ridge backwards; the concave section thus produced may detach from the main ridge and travel backwards as a *parabolic dune* (Figure 8.2). Travelling dunes often bury vegetation the remains of which are uncovered later. Whole woods may be overwhelmed and later reappear as 'ghost-forests'.

On coasts with a greater sand supply and gentler winds, dunes can form and become stable *in situ*, new dunes forming to seaward as the coast builds outwards. Neither moving nor stable dunes are necessarily aligned at 90° to wind direction but, probably due to geostrophic forces, are aligned at a somewhat lesser angle. Dune stability can result from vegetational succession, climatic change or coastal management. Stabilized dunes, sometimes referred to as fixed or grey dunes, as distinct from moving or yellow dunes, can persist for centuries. On the Dutch coast two main dune series are recognizable. One of these is pre-Roman and the other appears to have formed between the twelfth and seventeenth centuries. As dunes develop the organic content and nutrient level of the soil increases (that is, following an initial decline from tidal litter conditions). The first *Ammophila* dune 'soil' will have organic and nitrogen contents of less than 0.5 and 0.01% respectively. However, by the stage of grass-sward formation organic content exceeds 1%. Little is known about microfloral development in dunes, but some nitrogen fixation occurs and there is some

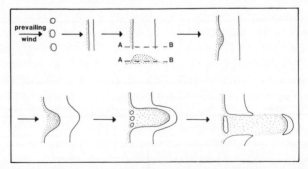

Figure 8.2   The formation and development of parabolic dunes. (After Ranwell, 1972.)

association of the bacteria with plant roots. This is certainly the case with some dune scrub species such as *Hippophae* (sea-buckthorn) which has root nodules. Fungal mycorrhizae are also important. Recent work on Australian dunes has shown that fungi are commonly associated with colonizing plants, whereas it had previously been thought that they were important only in older base-deficient dune soils (Jehne and Thompson, 1981).

The water-holding capacity of dune soils increases as the organic content builds up. There is a considerable shortage of water in young dunes and this limits the number of species able to take part in colonization. In most dunes the primary source of water is rain, but once it has percolated through the dune to the underlying water table it is unavailable to the dune ridge plants. Three main strategies have evolved in response to this problem. (1) Dune annuals undergo their vegetative cycle in autumn and spring or other wet seasons and survive the dry periods as seeds. Perennials either (2) produce a deep rooting system and are thus able to draw water from a greater vertical zone, or (3) produce an extensive shallow shower-exploiting root system. Water availability is enhanced by diurnal temperature change which leads to dew formation, the dunes acting as large water condensers.

Water percolation has two important chemical effects on dune soils. Dune sands often have an initially high calcareous content (i.e. are base-rich) due to shell fragments, but this is gradually leached out so that after 300–400 years the top 1 cm of soil is base-deficient. This affects the plant species composition and a visual sign of this process is colonization of the dune by species, such as heathers, characteristic of more acid soil. Nutrients are also washed out of the soil and enter the general water table. Rain itself contains small but significant amounts of nitrate and phosphate which will help support the dune vegetation, especially lichens which are able to absorb nutrients directly into their thalli.

As dune soils develop they become suitable for further plant species. There is an initial increase of diversity through the addition of various lichens, bryophytes, grasses and herbs. Some of these may participate in *hummock* formation. This is a characteristic feature of many dunes where local erosion occurs—small raised clumps or islands persisting because of their binding vegetation. After grass/herb layer formation, shrubby species such as buckthorn, rose and willow may become established. Some of these, for example *Salix repens* in Europe and *S. lasiolepus* in the United States, also help in hummock production. The final stage of natural dune succession is generally woodland or forest of, for example, *Abies, Pinus* or *Quercus*.

Dune slacks

Erosion of open sand surfaces may continue until the damp sand immediately above the water table is reached or until a layer of less erodable material such as shell or pebble is uncovered. Thus the hollows or slacks left after dune ridge formation tend to have moist or coarse floors. Erosion will not always proceed to such depths due to the sheltering effect of ridges or to sand redeposition. It is possible to distinguish three types of dune slack. These are (1) dry slacks which are generally grass-dominated; the water table lies 1 m or more below the surface and although there are a few deep-rooted species the majority of the vegetation is rain-dependent; (2) wet slacks in which the water table is shallow, moss growth is often extensive and blue-green prokaryotes such as *Nostoc* are common (these often provide good orchid habitat); (3) flooding slacks where the water table remains close to the surface in summer and covers the ground in wet seasons; here the vegetation is characterized by amphibious species such as *Polygonum* (knot-grass) or *Ranunculus*. Dune systems may develop a variety of more or less permanent water bodies. A recent study by Timms (1982) demonstrates decline in alkalinity and associated changes with distance from the foredunes.

Dune fauna

Dune vegetation provides a resource for terrestrial animals. Some dune meadows are grazed by domestic cattle but the dominant herbivore is often the rabbit. Grazing tends to maintain species diversity by providing a sward with some open areas for annual plants and by preventing scrub establishment. However, areas can be impoverished by overgrazing. A large range of invertebrates occurs in dune systems. Foredunes are the least suitable habitat because of their aridity and considerable temperature fluctuations (even in Britain the sand surface temperature may exceed 60°C). Some thick-shelled snails benefit from the high calcium availability and are able to escape the worst temperature stress by attaching to marram above the sand surface. The snail *Cepaea* has been the subject of several studies in population genetics. The insect community, particularly of beetles, wasps, flies and, consequently, the spider fauna is extensive. Actively hunting forms such as wolf-spiders predominate. Spiders in their turn are hunted by other wasps and by some vertebrates.

West coast deserts

These deserve brief mention here since they resemble enormous coastal dune systems. The major areas are the Atlantic Sahara and the Namib (with some 300 m-high dunes) in Africa, the Vizcaino and Atacama of Central and South America, and parts of western Australia. The ecology of these areas shows some similarity as illustrated by the Atacama and Namib deserts (Bramwell, 1980). A cold offshore northward current (the Humboldt and Benguela respectively) cools the low-level air, thus generating extensive fog. Over the coastal plain this air is heated, increasing its water-carrying capacity. Some condensation occurs with diurnal temperature decrease so that a very sparse xerophytic vegetation (for example *Tillandsia* and *Welwitschia*) is present. Further inland, at several hundred metres above sea-level, the fog layer lying between the lower cool air and the upper warm air sustains a fairly lush vegetation. Although the vegetation of the coastal plain is very sparse there is a rich population of invertebrates mostly sustained by the strandline or by windbore debris from inland. Forms such as ants and beetles are fed on by spiders and scorpions and these in turn by geckos, snakes or other vertebrate predators. The offshore currents lead to highly productive marine life which sustain rich fish, bird and mammal populations.

Shingle beaches

Shingle is a convenient word for sediment particles intermediate in size between sand and boulders. It forms similar coastal structures to those constructed of sand, for example long beaches, offshore bars and islands. Shingle spits tend to occur at corners where the coastline changes direction abruptly. They often have a recurved end due to wave refraction. Their intertidal zone is rather barren because of particle instability, but vegetation often occurs at or above the high-tide mark. Low nutrient and water availability add to the problem of instability, and those three factors, together with sediment composition, are important in determining the flora. Where conditions are minimal for plant survival an extensive lichen cover may develop; with additional resources various annual species—often those which are found on beach driftlines—are able to flourish. Perennial species occur where there is sufficient admixture of finer sediment and greater stability. Several rare plant species occurring in shingle are thought to survive here because of the lack of grazing pressure and competition. Raised shingle beaches in the Arctic, for example, may develop a characteristic low scrub and grassland including *Salix* and *Elymus*. Various prickly genera (e.g. *Rosa, Rubus, Ruscus* and *Ulex*) are common on raised shingle. In the most stable areas heathland may develop. Shingle provides important breeding sites for several birds including various terns.

Grassland and heath

Maritime grasslands do not necessarily show much difference from those inland. They usually contain some salt-tolerant plants and these may be important as a stock from which marshes, dunes and shingle areas can be colonized. Where marked differences occur they are mainly due to the climatic influence of the sea and/or to high ground inland which generally means milder and wetter coastal weather. Shell debris is often present, particularly on emerging and accreting coastlines, hence the flora may include many lime-loving plants, for example cowslip, harebell and common centaury in the British Isles. Similarly the grassland may be rich in snail species; for example, the heath or dune snail *Helicella italla* is found on inland chalk-down in southern England but further north is restricted to coastal areas. Various other invertebrate groups are common in maritime grassland; they include bushcrickets, grasshoppers, ants and lepidopterans. The *machair* of the Scottish Western Isles is a herb-rich coastal grassland initially accreted by marram succeeded by nitrogen-fixing trefoils (*Trifolium* and *Lotus* spp.).

Coastal heaths develop in areas with acidic or neutral soils and are often a late seral or zonal stage in dune development. They also occur on stable shingle and on cliff tops. Such heaths are dominated by ericaceous species, some of these being of limited coastal distribution (as is some of the other heath flora). In Britain the Cornish heath *Erica vagans* and early adder's tongue fern *Ophioglossum lusitanicum* are examples.

Coastal woodlands

Inland from heath, marsh or dune there is often a scrub zone followed by woodland or forest. This generally represents the natural climax vegetation of the area concerned. On the western seaboard of North America moving northwards, the rain forest of Central America is replaced in Lower and

Southern California by coastal scrub or *chaparral* (similar to the European Mediterranean *maquis*). Further north this is replaced by dry woodland and then by coniferous forest with Douglas fir, red cedar and spruce. On the eastern seaboard the coastal plain of Central Florida and South Carolina has a climax *Magnolia* forest with *Quercus virginiana* (live-oak) and other species. This is followed inland and northwards by pine forest and then by wet deciduous woodland. The eastern Canadian coast has coniferous forests. In subtropical areas west and/or windward coasts develop a maritime rain forest similar to equatorial forest. Such areas are found for example in Central America, the Caribbean, the Philippines and Indo-China. In the latter case, beach shoreline vegetation characterized by *Ipomaea* gives way to wooded scrub or coastal forest with *Barringtonia* and *Casuarina* trees, which is succeeded inland by full rain forest. In north-eastern Australia rain forest is found in the north of Queensland but to the south gives way to temperate *Eucalyptus* forest. One of the most interesting wildlife areas in the world because of its extraordinary range of endemic species, for example of *Banksia* trees, is the corresponding *Eucalyptus* forest of Australia's south-west corner.

Cliffs

Cliffs are found on embayed and plains coasts. In the latter case the cliffs are likely to be low and to consist of relatively uncompacted material. They are liable to slipping and to rapid storm erosion. The fauna and flora therefore reflect the frequency of cliff-face destruction and will be derived mainly from the clifftop habitat. Some plants characteristic of disturbed soil, such as coltsfoot (*Tussilago farfara*—a European species also found in North America) and horse-tails (*Equisetum*) are likely to occur. More stable cliffs may develop a bog or marsh-like vegetation.

Rocky cliffs, especially when high and ledged, provide an extensive wildlife refuge and often have seabird colonies (Chapter 10). Such cliffs have rarely been studied due to access difficulties. The vegetation is zoned from the base to the clifftop partly as a result of salt tolerance. Above high-water-mark there is often a salt-tolerant lichen zone. In northern Europe the characteristic species in ascending order are *Verrucaria maura, Xanthoria parietina* and *Ramalina* sp.; they are followed by a zone of salt-tolerant angiosperms including *Armeria, Spergularia rupestris, Festuca rubra* and *Plantago maritima*. Sea-cliff plants have considerable desiccation problems, not only due to salt content of the soil (a problem shared with marsh, mangrove, and dune plants) but also due to increased transpiration resulting from wind. *P. maritima*, which is now found in the Antipodes as well as circumboreally, is one of several species reducing transpiration by having a hairy surface. Other plants counter such loss by adopting a rosette growth form or by reducing leaf area. Towards the upper cliff edge the species composition merges into the local clifftop vegetation. Only a few ferns and mosses are sufficiently salt-tolerant for sea-cliff life, but a few such as *Asplenium marinum* are characteristic of this habitat. Some of the plant (and probably animal) species which flourish in the rich soil built up by weathering, plant decay and bird droppings include forms which are rare elsewhere because of grazing and predation pressure.

Swamps and mires

Chapters 5 and 7 have already dealt with some special habitats produced at the interface of fresh and salt water. There are other maritime and shore habitats modified by extreme precipitation or retention of water. Swamps and mires contain some of the most highly regarded remaining natural wildlife habitats. Examples could be drawn from various delta areas, from the Florida Everglades and from the Camargue area of France. One of the best such areas in Europe is the Cota Doñana bordering the mouth of Spain's Guadalquivir River. The river channel is fringed with reeds and then by a transition zone of sedges and bulrush (with several higher islands) leading to inland scrub vegetation. In summer the marshes are very dry and the huge spring bird flocks of ducks, geese and waders move to lagoon areas, moving back on to the marsh with the autumn rain. There is a rich amphibian and reptile fauna including lizards and snakes.

In northern high latitudes, *tundra* (muskeg) is the dominant vegetation. It mainly consists of moss, lichen and sedge. The soil moisture is frozen for much of the year. Precipitation and evaporation are low. The summer thaw does not penetrate far into the soil, so the excess water is unable to drain away and a marbled pattern of vegetation and shallow water develops. South of this permafrost zone the soils are able to dry out in summer months so that peat may be unable to persist and low scrub may develop. However, close to the coastline, for example in much of British Columbia, Newfoundland and Nova Scotia, the rainfall remains high enough for peat survival and growth, hence the coastal land is covered by blanket bog. A similar situation holds in Scotland, Ireland and Norway. Tundra or peat mires are rare in the Southern Hemisphere simply because there is little land in the appropriate latitudes; such habitats occur in the Falklands, South Georgia and other South Atlantic islands.

The ice edge

The area of the two great polar ice caps varies extensively with the season. In the Arctic the Siberian, Alaskan, Canadian and much of the Greenland coasts are ice-bound during the winter, but only part of the northern coasts of Greenland and Ellesmere Island are permanently in pack ice. The polar bear and walrus are endemic Arctic species usually found near the ice edge. In the Antarctic, the ice retreats towards the underlying land mass. The coastal vegetation is even sparser than that of the Arctic. There are about 70 species of moss and 15 of liverwort, forming a thin and patchy peat cover. Higher plants are absent except on the Graham Land Peninsula islands where *Deschampia antarctica* (tussock grass) and *Colobanthus quitenuis* (cushion plant) form small vegetation mounds. The vegetation helps support a few mites and insects (e.g. midges and springtails) and there is a limited soil fauna with mites, springtails and nematodes. The seaward edge of the coastal strip is very important as a breeding site for penguins and seals. The four truly Antarctic seals are the fish-eating Weddel and Ross seals, the crab-eater and the leopard seal. The latter is a voracious predator, particularly of penguins. In general these seals are larger but shorter-lived than their Arctic counterparts; this is presumably linked to the greater productivity of the southern waters. The seal and bird colonies lead to considerable nutrient transfer between sea and land. Figure 8.3 shows feeding relationships in the Antarctic pack-ice zone.

Arctic and Antarctic coastal marine animals have developed many special physiological adaptations to cold water (see Clarke, 1983). The birds and mammals show diving adaptations and, in addition to thermal protection by fur, feather and fat, have blood-system modifications such as the counterflow in penguins' feet. The mechanisms of freezing resistance in invertebrates are not well studied; however, since their body fluids are at least iso-osmotic with seawater they will not freeze unless the sea freezes first. It is thought that freezing damage is avoided, as in boreal molluscs, by withdrawal of water to freeze extracellularly. In polar fish, freezing is avoided either by supercooling (which is important since the blood is hypo-osmotic) or by possession of antifreeze molecules of peptide or of glycoprotein. Growth rates, oxygen consumption and the proportion of metabolism devoted to reproduction appear to be lower in polar than in

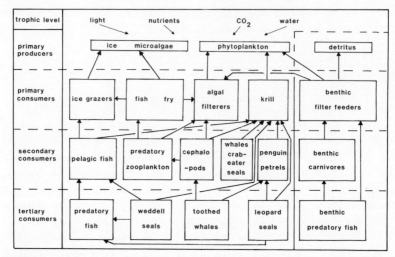

Figure 8.3 Food chain relationships of the Antarctic pack-ice zone. (After Knox in Holdgate, 1970.)

temperate forms. It has been argued cogently (Clarke, 1983) that this is a response to the extreme seasonality of primary production in high latitudes. There is an interesting community of diatoms and other forms adapted to live on the underside of sea ice. Polar mammals and birds provide extreme examples of the geographical linkage of different coastal areas through migration—the Arctic Tern, *Sterna macrura*, nests in the Arctic summer and then flies south for the Antarctic summer, and some fur seals migrate from the Pribiloff Islands (Bering Sea) to breeding grounds off California.

CHAPTER NINE

THE INSHORE HARVEST

Not many years ago the increasing effort and ingenuity applied to sea fisheries meant that the marine harvest kept pace with the human population increase. However, in the 1970s the rate of fisheries expansion slowed so that the amount of fish caught per person began to decline (Wheaton, 1977). Estimates of the maximum sustainable yield of food from

Table 9.1 Catch size (tonnes) of the thirty marine species which each contributed more than 400 000 metric tonnes to the total 1982 fish catch of 76 772 800 t (FAO, 1984).

Alaska pollack	*Theragra chalcogramma*	4 474 744
Japanese pilchard (= sardine)	*Sardinops melanosticta*	3 966 090
Chilean pilchard (= sardine)	*Sardinops sagax*	3 252 063
Atlantic cod	*Gadus morhua*	2 264 251
Chilean jack mackerel	*Trachurus murphyii*	2 195 387
Capelin	*Mallotus villosus*	1 852 444
Anchoveta (= Peruvian anchovy)	*Engraulis ringens*	1 831 627
Chub mackerel	*Scomber japonicus*	1 591 964
Atlantic herring	*Clupea harengus*	984 264
European sardine (= pilchard)	*Sardina pilchardus*	896 975
Gulf menhaden	*Brevoortia patronus*	854 336
Largehead hairtail	*Trichiurus lepturus*	794 213
Skipjack tuna	*Katsuwonus pelamis*	762 314
European anchovy	*Engraulis encrasicholos*	684 575
Cape horse mackerel	*Trachurus capensis*	668 374
Atlantic mackerel	*Scomber scombrus*	629 541
Pacific thread herring	*Opisthonema libertate*	572 798
Blue whiting	*Micromesistius poutassou*	552 730
Yellowfin tuna	*Thunnus albacares*	535 725
Norway pout	*Trisopterus esmarkii*	531 986
Antarctic krill	*Euphausia superba*	530 003
Sprat	*Sprattus sprattus*	525 135
Californian pilchard (= sardine)	*Sardinops caerulea*	512 995
Saithe (= pollock)	*Pollachius virens*	510 673
Atlantic redfishes	*Sebastes spp*	489 460
Pacific cupped oyster	*Crassostrea gigas*	484 842
Cape hakes	*Merluccius capensis, M. paradox*	445 481
Clupeonella	*Clupeonella delicatula*	434 655
Haddock	*Melanogrammus aeglefinus*	423 497
Atlantic menhaden	*Brevoortia tyrannus*	400 341

the sea vary but it is generally thought that 120 million metric tonnes is possible. It has been suggested that increased exploitation of squid (say $10 \times 10^6 \, t \, y^{-1}$) and of krill ($100 \times 10^6 \, t \, y^{-1}$) might double this yield. Krill first enters the UN Food and Agricultural Organization's statistics in 1975 and within six years was ranking as the twenty-fifth most important fishery resource (Table 9.1). This chapter does not set out to be a potted fishery text but hopes to promote interest in higher vertebrates as well as fish, harvested invertebrates (including shrimp) and algae in coastal waters.

Sponges

Although some sponges are known to produce ergosterols and related substances of medical interest, commerce is concerned with two other main aspects. The first is that some clinoid sponges are pests because they bore into the shells of oysters, scallops and other commercial species. This is accomplished by acid secretion, and leads to risk of other infections and detracts from saleability. The second is harvesting for domestic and industrial use as sponge—the properties of which are dependent on the spongin skeletal fibre system. Most sponges are taken in shallow water by diving, by rake or by rope-looped pole. The industry has suffered from overcollection and from competition from plastic products. However there has been renewed interest in this renewable resource, and world annual production, mainly from Spain, Turkey and Tunisia, is over 250 000 kg dry weight (Firth, 1969).

Corals

The precious coral *Corallium rubrum* is collected from subtidal rock in the Mediterranean and off Japan, but stocks have declined in many areas. It is a member of the soft-corals, the Alcyonaceae, but has a calcified skeleton. There is a far more extensive trade in true corals (Scleractinia or Madreporaria) for marine curios or for aquaria. In some areas, particularly where explosives have been used, this has led to reef destruction or impoverishment and to increased erosion.

Worms

Nearly all general marine invertebrate texts remark on the precise annual spawning of the Pacific palolo worm and the associated Polynesian feast (the released hind ends are made into soup). The main worm harvest is more prosaic, consisting of various species collected for bait. This is mainly for sea angling. In the United States national surveys in the 1960s estimated that there were 8.3×10^6 regular coastal anglers who between them caught 380×10^6 kg of fish each year. The bait included at least 30×10^6 bloodworms (*Glycera dibranchiata*) and a similar number of sandworms (*Nereis virens*) which were dug commercially and valued at $1 m. Some areas may be dug 20 times in a year. Overall production does not appear to have suffered as a result of such digging intensity, but the average size of worms has declined. In Europe the chief bait species are *Arenicola marina* (lugworm) and *Nereis* spp. (ragworm), and some concern has been expressed over local disturbance of habitat by digging, especially in *Zostera* beds.

Crustacea

The world's major fishing catches are shown in Table 9.1. It is estimated that crustaceans comprise between 1–2% of the total. This proportion varies considerably between countries, as does that of the various crustacean types. In Great Britain crustaceans comprise about 5% of the total catch value, with *Nephrops norvegicus, Homarus vulgaris* and *Cancer pagurus* being the most important species. In the United States the crustacean catch is about 20% of the landed weight and about 40% by value with shrimp, crabs and lobsters dominating.

Shrimps

On a world basis shrimps (or prawns — there is no internationally accepted distinction) comprise nearly two-thirds of the crustacean catch excluding krill. *Penaeid prawns* are important in the main fisheries. These are found in the subtropics where extensive mud at 10–50 fathoms provides adult feeding grounds and more estuarine inshore areas provide nursery grounds. The most important areas are the west coast of India and Pakistan, the coastal waters of Japan, China and the Indo-Pacific, the Gulf of Mexico, and off north-east South America. Shrimp are caught mainly by otter trawl. The aquacultural potential of penaeids has attracted much interest since they are less aggressive and have faster growth rates than *caridean prawns*. There are other differences between the groups; for example, in carideans the eggs are carried by the female until hatching at the first larval stage (nauplius), but in penaeids the eggs are shed directly into the sea. In both cases there is a complex series of moults to pass through the naupliar, protozoeal, zoeal and mysid stages. Most aquacultural work has been undertaken in Japan with *Penaeus japonicus*, which was the first prawn species to be reared commercially through its entire life cycle. Similar work is being undertaken on the Pacific *P. orientalis* and *P. monodon* as well as on the brown, pink, spotted pink and white shrimp of the Gulf of Mexico.

The European shrimp fisheries are for caridean species such as *Pandulus borealis* and *P. montagui* which is found closer inshore. Many shrimp species are protandrous hermaphrodites, that is, they first mature as males but later become females. In *Pandulus*, a proportion—which is larger in more southerly populations—develops directly into females; sex change also occurs later in southern individuals. Prawns are fairly voracious predators. *P. montagui* feeds on polychaetes including *Sabellaria* and *Pectinaria* as well as on amphipods. Tiews (1969) notes that in the extensive brown shrimp (*Crangon*) fishery off Germany more than four times as many shrimp are taken by natural predators as by man. Prawns in turn form a large part of the diet of inshore fish such as rockling, smelt, rays, flatfish and various gadoids. One caridean genus *Periclimenes* has specialized in forming cleaning associations, particularly with animals such as anemones, jellyfish, sea slugs and echinoderms which afford some protection from predation, and with fish.

Crabs

Two distantly related decapod groups are commonly referred to as crabs—the Brachyura or true crabs and the Anomura comprising squat lobsters and hermit, porcelain and lithodid crabs. The lithodid *Paralithodes camtschatica*, sometimes called the king crab (but not to be confused with *Limulus*, the king or horseshoe crab, which is not even a crustacean), is the basis of an extensive fishery centred around Kodiak Island, Alaska. Female king crabs lay upward of 150 000 eggs just before moulting and recopulating in early spring. Larvae settle from the plankton after up to 12 weeks, but the young crabs remain in huge benthic clusters (pods) for 3 to 4 years before becoming more solitary and breeding in their fifth year.

The true crabs include many fished species. Some, such as the red crab *Geryon quinquedens*, are shelf-edge rather than strictly coastal species. Several of the latter belong to the genus *Cancer*. This includes the west

North American *C. magister* (the Dungeness crab) and *C. pagurus*. About 15 000–20 000 t of the latter are landed in Europe each year. The majority are caught in baited pots but it is unusual for berried (i.e. egg-carrying) females to be caught; the reasons are not known, but nevertheless fortunate both for the crabs and for the fishery as a self-controlled conservation mechanism. Copulation takes place immediately after the female moults (often in August) but before the shell hardens. After copulation a 'sperm-plug' is produced from spermathecal fluid which prevents loss of sperm from the distended vulva. Spawning, in which the eggs are laid but immediately attach to the females' pleopods (feathery abdominal legs), occurs several months after impregnation. The eggs, 1.5–3 million per female, are carried for several months before hatching as protozoea in spring to late summer. After about 3–4 weeks and a series of moults in the plankton, the larvae settle to start benthic life. *Cancer pagurus* is a vigorous predator and scavenger. It can crush fairly large mussels and whelks in addition to small sea-urchins: it also digs for burrowing bivalves (Edwards, 1979). Several crabs are known to undergo short migrations; some of these are of an onshore–offshore nature and associated with the breeding cycle. Recently there has been considerable expansion of the European fishery for *Maia squinado*. Considerable information about the migration and clumping behaviour of this species has come from sub-aqua observation. The blue crab *Callinectes sapidus*, which has a major predatory influence on benthic community composition, is itself heavily predated by man on the United States eastern seaboard.

Lobsters and spiny lobsters

Three species of true lobster (with a large central rostrum above the eyes and with at least small pincers on all thoracic legs) are found in North Atlantic coastal waters. They are the American and European lobsters (*Homarus americanus* and *H. vulgaris*) and *Nephrops norvegicus*. The former two species were regarded generally as inshore species and mainly fished with baited pots. However, both species extend further offshore than was originally thought. *H. americanus*, in addition to its rocky coast habitat, burrows into mud and grows to a much larger size in shelf canyon areas than in the intensively fished inshore zone. Studies of this species in Nova Scotia (Wharton and Mann, 1981) show that its predation on sea-urchins reduces the latters' grazing pressure on kelp, which in turn provides more shelter for lobsters and also enhances secondary productivity to the lobster's benefit.

Nephrops norvegicus, variously known as Dublin Bay prawn, Norwegian lobster and scampi, is a characteristic inhabitant of mud off northern European coasts. It constructs burrows which may have several chambers and entrances and may be shared with other species. *Nephrops* tends to be night- or dawn- and dusk-active. Females in berry often live nearer the burrow exit than the males; this is thought to be because a high level of oxygen is required for egg development.

Spiny lobsters (also called crawfish, crayfish or rock lobsters) have a pair of rostral spines and lack pincers on the first four pairs of walking legs. Three genera (*Jasus, Palinurus* and *Panilurus*) are important in fisheries. *Palinurus* includes *P. elephas* of Europe and *P. mauritanus* of the west African coast. The former species is mostly caught by pots, but the latter by trawl and by bottom tangle nets. There are numerous *Panilurus* species in tropical and subtropical coastal waters. *P. cygnus* is the basis of the Australian west coast cray fishery. *Jasus* is a southern temperate genus thought to be derived from a common circumpolar ancestor. *J. lalandii* is fished off south-west Africa where it is a major component of the kelp-bed ecosystem, feeding on the ribbed mussel *Aulocomya* and other species. As far as is known all spiny lobsters have a very long planktonic phase lasting up to 11 months; the final larval stage is the leaf-like *phyllosoma* and a product of about 30 moults. It therefore seems unlikely that spiny lobsters could be cultured commercially.

Mollusca

Three molluscan groups, that is bivalves, gastropods and squid, have a high fisheries value and could probably be exploited further, the first and possibly the second by culture and the third by extension of fishing areas and methods. A few species are of commercial value through the souvenir or other trades; capiz shell from the Philippines is a good example.

Bivalves

The bivalves include the scallops, clams, oysters and mussels. Most of these forms are sessile when adult, but several scallop species can swim actively over short distances. This appears to be mainly a response to the presence of predators. Thus the queenie *Chlamys opercularis* can be induced to swim in aquaria by adding a few drops of sea water which have been 'contaminated' by the starfish *Asterias rubens*. Oysters are characteristic inhabitants of intertidal and shallow waters in all except polar regions. The two most important genera commercially are *Ostrea* and *Crassostrea*. The shell valves are unequal, as in scallops, and sediment-dwelling forms live with the convex (left) shell resting on the substratum. In *Crassostrea*, which includes the commercial American, Portuguese and Pacific oysters (*C. virginica, angulata* and *gigas*), eggs and sperm are shed into the sea. In *Ostrea*, which is hermaphroditic, sperm is taken in with feeding currents and fertilizes eggs in the mantle cavity; the developing eggs remain here for 1–2 weeks before being shed into the plankton. Larvae settle on to suitable substratum (*cultch*) which in the oyster industry is often crushed shell or sometimes tiles or bundles of twigs. Newly settled larvae are termed *spat* (because it was thought that *Ostrea* spat out its young). In most culture systems the spat are carefully removed from the cultch and then relaid as seed oysters in hanging nets (lantern culture) or on trays stacked below rafts. In many areas natural production of spat is unreliable, hence oyster hatcheries have been established for spat and seed production. Oyster farming has been beset by misfortune arising from accidental introduction of parasites, pests and competitors when oysters have been transferred from area to area. Internal or shell infections are caused by various protistans such as *Hexamita* and *Minchinia* and by fungi. Predators include the oyster drill *Urosalpinx cinerea*, originally an American Atlantic coast species and now found in the Pacific and on European coasts. Competitors for settlement space and for planktonic food include the Australian barnacle *Elminius modestus* and the American slipper limpet *Crepidula fornicata*; both are now common on some sheltered European shores.

Various burrowing clams are the basis of local fisheries. Some, such as *Mya arenaria* (the soft-shell clam or sandgaper), which is highly prized in the United States but little used in Europe, are of considerable value. Some progress has been made with artificial rearing of clam species. The hard-shell clam *Mercenaria mercenaria* has been experimentally introduced to some European estuaries.

Mussels, in particular *Mytilus edulis, M. galloprovincialis* and *M. californianus*, are probably the most thoroughly studied of all marine invertebrates and a vast literature exists on topics ranging from isozyme systematics and ultrastructure to community structure and industrial utilization. This interest has arisen partly from the ubiquity of *Mytilus* and related genera, and partly from their potential contribution to world protein supply. Production of mussel meat of up to $2000 \, \text{kg ha}^{-1} \text{y}^{-1}$ (wet weight) has been reported from Europe. This is higher than grassland beef production. Culture of mussels usually involves collecting of spat or seed mussels, by relaying from natural populations, by collection on wooden stakes or by hanging fibrous ropes in suitable areas, afterwards thinning out to other ropes to help adult growth. The latter method seems the most promising and also helps to counteract predation by benthic forms such as starfish. *Mytilus* and related genera such as *Modiolus* are dominant organisms in various habitats (see Chapters 3 and 4). In a Baltic inshore community it has been estimated that mussel beds filter $56\,g\,C\,m^{-2}\,y^{-1}$, which is equivalent to about one-third of the overlying phytoplanktonic production. Much of the carbon gained is channelled into growth and reproduction but a large proportion is lost in faeces which form an important resource for bacteria and thus for detritivores. *Mytilus* larvae are important as food for zooplankton, including herring larvae, and mussel populations are important in nitrogen and phosphorus recycling (Kautsky and Wallentius, 1980).

Gastropods

The gastropods comprise limpets, snails and allied forms; a few species form the bases of local, mainly small, fisheries. Among these are the ormers or abalones, conches, whelks and winkles. Winkles for example are the twelfth most valuable species in Northern Ireland's fishery which is otherwise dominated by whitefish, herring, squid and Dublin Bay prawn. Conches are fished commercially for food mainly in the Caribbean and in the Indian Ocean; in addition a large number of shells of conches and other gastropods, for example *Conus* spp., enter the marine souvenir trade. Abalone (*Haliotus* spp.) is fished off various tropical and temperate coasts including south and west Australia, South Africa, Japan and north-west America. Abalones, like winkles, are algal grazers. At present in Europe abalone has the highest price to weight ratio of any marine food. Some success has been achieved in abalone farming.

Squid and octopus

By weight the world squid catch approximates to that of shrimp and (excluding mariculture of oysters and mussels) it is certainly the largest molluscan fishery. Much of the catch is from the open ocean but there is a substantial inshore component, for example in New Zealand, much of which is based on the genus *Loligo* which occurs over coastal sand bottoms. The Newfoundland squid fishery depends on migrating *Illex illecebrosus* which come close inshore in late summer. Squid fishing is usually by mechanical jigging with coloured barbless hooks. Squid feed mainly on fish and crustaceans and are preyed on by elephant seals, sea-lions, toothed whales and some penguins and other sea-birds.

Octopi are predominantly coastal night-active predators of other molluscs, crustaceans and fish; they are taken in some local fisheries, often as a by-catch. Like squid they have a venomous bite which contains a neuromuscular toxin; in the Indo-Pacific blue-ringed octopus, which only reaches 10 cm length, this is strong enough to kill humans. Octopi produce salivary enzymes which partly digest food before it is taken into the body. Much work on the function of nervous systems has been conducted on *Octopus vulgaris*.

Commercial fish

About 3% of the world's food supply comes from the sea. The seaweed harvest of about 3×10^6 t wet weight per annum is negligible in comparison with the fish catch of 70×10^6 t. Table 9.2 shows the dominance of the northern and central western Pacific and the northern Atlantic in producing this catch. The reasons for this are twofold. Firstly, these areas contain at least one-half of the world's continental shelf, and secondly, these areas are more heavily fished. The latter statement is to be treated with caution since fisheries are still developing rapidly in some other areas. However since heavier fishing may be counter-productive and better management may be introduced it is possible that northern areas will continue to dominate world fisheries. The greater part of fisheries

Table 9.2 Total landings of fish and shellfish (million tonnes) from major waters. (Based on FAO, 1984).

Area	1960	1970	1981
North Atlantic	9.8	14.9	14.5
Central and South Atlantic	4.3	6.8	8.7
Mediterranean	0.8	1.1	1.7
Indian Ocean	1.8	2.5	3.5
North and Central West Pacific	12.4	19.8	27.9
Other Pacific areas	4.5	14.9	9.9
Southern oceans	0	0	0.6
Total marine	33.6	61.0	66.7
Total inland waters	4.2	5.3	8.1
Total	37.8	66.6	74.8

production can be assigned to three types of fish which, as far as 'western world' households are concerned, tend to come in three packaging shapes.

Demersal fish

Demersal or white fish are basically fish of the benthos which are usually marketed in oblong cardboard packages as 'breaded portions' or fish fingers. The main species concerned are gadoids, for example pollock, cod, whiting, haddock, pout and hake, or the flatfish such as plaice, flounder, sole, dab and halibut (however the latter group only comprise 9–10% of the total). The North Atlantic cod (*Gadus morhua*) fishery shows a history common in un- or mis-managed stocks. The species has been fished in European inshore waters for centuries but with growing human population

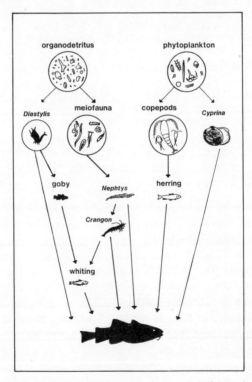

Figure 9.1 The major food chain relationships of the cod *Gadus morhua* in the Western Baltic. (Based on Arntz, 1978.)

the supply became insufficient to meet demand. Boats were made larger and sought coastal grounds further afield. Thus cod from Iceland, Labrador and Newfoundland became important by the sixteenth century. The fishery also moved offshore into deeper water. Supply from distant and deeper waters was aided by the advent of steam power and refrigeration. By 1920 the North Sea cod stock had been exploited as far as, or beyond, its maximum yield. The same had happened to the north-east Atlantic stock off Iceland by 1930 and to north-west Atlantic and Barents Sea stocks between 1950 and 1960. The usual response to a declining resource, apart from raising prices, has been to increase fishing effort or to seek alternative species. Thus fisheries for Alaskan pollock, blue whiting, Norway pout and hake have built up as cod stocks have declined. Feeding relationships of cod are shown in Figure 9.1. Large cod reach 50 kg but individuals of 5–10 kg are regarded as of good size. Such females produce up to 9×10^6 eggs which, as in most fish, are planktonic.

Pelagic grazers

Fish packed in oval or oblong tins include anchovy, herring and sardines. These are species which form extensive shoals and feed by filtering phytoplankton and/or zooplankton. They include several forms such as the menhaden (*Brevoortia*) which are used extensively in fishmeal and oil production. Like many other fish, young *Brevoortia* are more coastal than when adult. Juvenile *B. tyrannus* are important ecosystem components in inshore areas with high planktonic productivity such as the Chesapeake and Narangasett Bays. Upwelling areas also support large numbers of filter-feeding fish, obvious examples being the anchoveta *Engraulis mordax* off Peru and *Sardina* or *Sardinops* species off west Africa and California.

Some of the most studied fish stocks are of the North Atlantic herring *Clupea harengus*. The Domesday Book (1086) lists rents for herring fisheries off the English coast. The history of the Baltic and North Sea stocks is related to the foundation of European fishing fleets and hence to naval power, as well as to the growth or decline of various trading ports, for example the Hanseatic League towns. After the Baltic herring fishery collapsed rivalry between the English and Dutch over North Sea stocks was involved in the wars of 1652, 1655–7 and 1670–4. The herring is a fickle fish whose recruitment is unpredictable, but it is certain that North Sea stocks were overfished by 1950 and the Atlanto-Scandinavian stock (north of Norway) by 1955.

A replacement resource was found by exploiting stocks of the capelin *Mallotus villosus*, a boreo-arctic species which feeds mainly on euphausiids and *Calanus*. Like the herring, the capelin is bottom-spawning but it prefers pebble and gravel areas with a mean particle size between 0.5–2.5 cm. Some stocks spawn on beaches, others in depths down to 100 m. Beach-spawning capelin rely on wave uprush to carry them ashore but then orientate to face landward against the backwash. They come ashore in large shoals but usually spawn in groups of three—one female between two male fish. The female lays up to 60 000 eggs in a scooped-out hollow. The eggs adhere to the pebbles and hatch from a week to a month later; the length of time is dependent on the temperature. Between the early 1960s and late 1970s capelin catch rose from 5×10^4 to 3×10^6 t and since then has remained among the five most important world fisheries.

Pelagic predators

Pilchard and sardines are rather more selective and predatory feeders than anchovies, but not as selective as jack mackerel (*Trachurus* and *Caranx*) and mackerel (*Scomber*), which predate the larger zooplankton. *S. scombrus*, the Atlantic mackerel, for example, takes copepods, young herring and sand eel. In the 1960s there was a considerable increase in fishing of this mackerel and of its Pacific relative *S. japonicus*. Larger predatory bony-fish such as bonito and tuna species are mainly caught in oceanic waters. They, like mackerel, are packed in round tins, but this tells us less of their ecology than does the dark flesh colour which results from the extensive blood supply to the musculature which is needed to produce the fast swimming necessary to catch other fish.

Fisheries research

Much fisheries research and management is aimed at the goal of *maximum sustainable yield*; that is the largest total weight of the species concerned which can be caught annually without causing a decline in subsequent years (see Cushing, 1968 or Harden Jones, 1974). Basically this is a matter of seeing that (1) the combined natural and fishing mortality is balanced by the rate at which young fish join the fished population (recruitment) and the rate at which the population gains weight, and that (2) this balance is achieved at a level which allows efficient growth and reproduction to have occurred (Figure 9.2). There are two types of overexploitation leading to

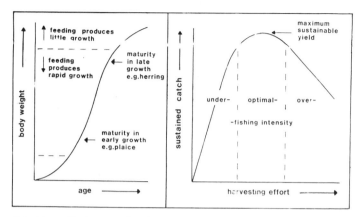

Figure 9.2 Generalized curves for the growth of marine fish and for yield/intensity in fisheries.

reduced catch weight, namely *growth overfishing* and *recruitment overfishing*. In the first, the catch declines because fish are being removed from the stock before their efficient growth phase is complete. Thus (Figure 9.3) the weight of a 500 g fish taken from species *A* would have doubled within two years if it had been left in the fishery. Recruitment overfishing is even more serious because the fishing intensity is such that insufficient breeding stock is left to supply recruits. This for example would happen to species *B* (Figure 9.3) if the bulk of fish being taken were of 400 g or more (4–5 years or older), since very few mature fish would be left.

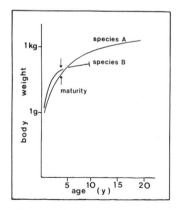

Figure 9.3 Maturity in relation to maximum growth in two species of fish. See text. (After Cushing, 1968.)

The growth curve of species *B* approximates to that of the North Sea herring (and species *A* to the plaice). Clupeoids are prone to recruitment overfishing since they mature near their growth/age asymptote. Herring also illustrate the fisheries problem of *stock identification*. Thus North Sea herring caught on a feeding ground may be comprised of fish from three different stocks—the Buchan, Dogger and Downs groups, so called from their separate spawning areas. It is often difficult to distinguish between different stocks of the same species and it is usually necessary to resort to statistical analysis of the variation in the number of vertebrae, fin rays or age/size increment (such statistically determined parameters are termed *meristic characters*). These herring stocks also illustrate the 'circuit of migration' displayed by many fish. This comprises adult migrations between the feeding and spawning grounds plus passive drift of larvae from the spawning to nursery areas followed by juvenile migration to join adult stocks (Figure 9.4).

Very few studies have attempted to view fisheries as a multiple-species resource although it is obvious that the success or failure of one fishery or species frequently influences that of another (May *et al.*, 1979). Particular problems arise, for example when the net mesh size suitable for allowing reasonable survival of one species takes large numbers of juveniles of another species. The capture of whiting in *Nephrops* nets or of whiting and

Figure 9.4 Adult migrations (solid lines), spawning areas (letters) larval drift (dashed) and juvenile migrations (dotted) of the herring *Clupea harengus* in the North Sea. (After Cushing, 1968.)

haddock in sand-eel fisheries illustrate this difficulty. Clearly the relationship between krill and baleen whale populations is another such 'fisheries ecosystem' problem.

Fish culture

Coastal fish have been the subject of aquaculture for many years. Two main examples are milkfish (*Chanos*) in the Far East and mullet (*Mugil*) in southern Europe. Both are cultured traditionally in coastal lagoons or ponds with high primary productivity which are stocked by flooding with sea water containing the larval fish. However nowadays it is common for *Chanos chanos* juveniles to be netted offshore and then transferred to the culture ponds (tambaks). These are usually situated just above high-water mark and initially flooded to about 10 cm depth. The ponds are fertilized with natural material such as mangrove leaves, grain or animal dung, or with artificial fertilizer. Salinity is controlled (10–20‰) or left to change naturally. The tambak floor develops dense populations of prokaryotes and protists which are fed on by the fish. Considerable success has been achieved by evolving a polyculture system in which the red alga *Gracilaria* (see later) is grown in addition to prawns and milkfish. *Chanos* production in these systems can exceed $1500 \, \text{kg ha}^{-1} \, \text{y}^{-1}$.

Recently there has been considerable expansion in rainbow trout and salmon farming. It has also proved possible to rear and to hybridize various flatfish under laboratory conditions. Pure strains of some flatfish have been produced by inducing apomixis—that is development of eggs without fertilization by sperm. Such techniques mirror the terrestrial shift from hunting to husbandry which first occurred several millenia ago.

Other coastal fish

Fish play an important part in most intertidal and inshore communities. There are so many species and adaptations to the coastal habitat that it is impossible to treat the subject adequately here. The last section mentioned just a few of the commercial species although it has been estimated that nearly 70% of these are dependent on estuaries as nursery areas. Work in Japan has suggested that decline of several commercial species such as sea-bass and sea-bream was associated with decline of *Zostera* beds (Kikuchi, 1974).

Herbivorous fish control the biomass of non-symbiotic algae in coral

F

reefs but represent only 10–30% of the total fish density, which ranges from 200–2000 kg ha^{-1}. In such complex communities fish have adopted a variety of bizarre roles and habitats, including various shelter and cleaning associations (Chapter 6). Beds of giant kelp may house up to 40 different fish species, among them the kelp-fish *Heterostichus rostratus* which also occurs among other seaweeds and eelgrass, its colour matching its surroundings. *Heterostichus* is a perciiform fish with an elongated body. Attenuation is common in boulder-shore fish as well as in sediment burrowers.

The blennies and gobies are characteristic coastal fish although a few species occur in other habitats. The butterfish *Centronotus* (= *Pholis*) *gunnellus* is a common North Atlantic intertidal and subtidal species and, as in many other shore species, after pairing and egg-laying, the male guards the eggs. *Centronotus* eats various small invertebrates and is a dietary component of herons and some other shore-feeding birds. The butterfish has an oesophageal blood plexus which aids in aerial respiration. Several intertidal fish have been shown to have endogenous activity cycles; these probably increase physiological efficiency in this cyclically changing environment. Other common shore adaptations are cryptic coloration, small size, desiccation tolerance and negative buoyancy (Gibson, 1982). The syngnathiform fish which include the pipe-fish, sea-horses and sea-dragons are also typically coastal but mainly associated with seaweed and sea-grasses in temperate and tropical waters. The sticklebacks are found in marine and fresh water in the Northern Hemisphere, and the genus *Gasterosteus* has been the subject of some classical behavioural studies.

Some fish species are characteristic of the lower reaches of rivers and hence could be regarded as coastal. Many are typical species of muddy slow-moving water; they include various suckers, carp and catfish. Other fish migrate through coastal areas during their life history. *Anadromous* species are those which are adult in the sea but breed in fresh water; examples are salmon, shad and striped bream. Several important fisheries, especially in the North Pacific, are based on salmon. *Catadromous* fish live in fresh water but breed in the sea; the best-known example is the eel *Anguilla*, but there are others, such as some of the New Zealand gobies.

Higher marine vertebrates

Various frogs and toads are predators in coastal swamps, marshes and mangals but no present-day amphibians are truly marine. However, several reptilian groups have marine or estuarine representatives. Two of these of considerable zoological interest are the sea-snakes and the iguanas.

Snakes and lizards

There are about 50 sea-snake species, all of which belong to the Elapidae—the family which includes cobras, kraits and mambas. Sea-snakes are more or less confined to the eastern Indian and the western Pacific Oceans. One species (*Pelamis platurus*) is found in the open ocean but ranges from the west coast of Panama to the east of Africa. It feeds on pelagic fish. *Microcephalophis* feeds on coral-reef eels. The sea-snakes show varying degrees of adaptation for marine life; some can remain submerged for two hours. They are mainly benthic feeders but tend to remain on the surface at night. The toxin of *Enhydrina* is the most poisonous snake venom known. Most species bear live young at sea.

One marine iguana inhabits rocky shores in the Galapagos and feeds on seaweed. The species has salt-secreting nasal glands. Eggs are laid in sandy areas on land.

Crocodiles and turtles

The American alligator is found in coastal areas but is not as truly marine as is the estuarine crocodile *Crocodylus porosus* of southern India, Indonesia and Australia. This crocodile has been greatly reduced in numbers because its hide is the most highly valued for the shoe and handbag trade; it is hoped that recent strong conservation measures may lead to recovery, as happened with the alligator.

There are seven species of turtle. The largest is the leatherback which is mostly oceanic, feeding on jellyfish. The other turtles have hard horny shields which are valued, especially in the hawksbill, for production of tortoiseshell. The hawksbill feeds on crustaceans and molluscs in rocky and coral areas. The green turtle has been hunted for eggs and meat; turtle soup was a traditional course at the Lord Mayor of London's annual banquet. This turtle is mainly a sea-grass herbivore but takes some invertebrates. Most turtles migrate to the beach of their birth to reproduce. The adults and eggs are thus particularly vulnerable to local hunting by humans and the young to natural predation as they cross the beach to enter the sea.

Marine mammals

Two members of the Carnivora, the polar-bear and the sea-otter, are more or less marine. The latter is a North Pacific species now re-introduced to some areas, for example Oregon and California, where it had become extinct, probably due to fur hunting. These otters feed predominantly on sea-urchins which otherwise would have a dominating grazing effect on kelp (Dayton, 1975). The resulting canopy increase favours lobsters in some way, but these are also eaten by sea-otters. It is possible that otter predation on urchins helped maintain the high algal productivity which was cropped by the now extinct Steller's sea-cow. The latter was the only temperate member of the Sirenia, an order now comprising the manatees (3 species, one of which is marine) and the dugong. These and the marine iguana are the world's only large herbivores feeding on macroalgae. It is possible they are responsible for the mermaid legend. The remaining marine mammals belong either to the Cetacea, that is whales, or to the Pinnipedia (seals, sea-lions and walruses).

There are 56 species of *toothed whales* and their smaller relatives the dolphins and porpoises. They feed on squid, crustaceans and fish. A few species inhabit rivers. An example is the Ganges dolphin or susu, which is blind and thus even more dependent on echo-location than are marine species. It is probable that all cetaceans are social animals. The Arctic White Whale is one of the species found closest to shore and is particularly vocal, hence its old whaling nickname of 'sea canary'. There are 10 species of *baleen whales* which feed by straining zooplankton through fringed horny plates (baleen) hanging from the upper jaw. Baleen was once used for stiffening corsets, but the chief whaling products were (and are) protein and oil. Baleen whales include the right whales such as the Greenland Right (so called because commercially they were the right whale to catch), the Grey Whale of the North Pacific and rorquals. The smallest and most coastal of these is the Minke or Pike Whale *Balaenoptera acutorostrata* (the specific name refers to its pointed snout). The management of whale fisheries has been attempted by the International Whaling Commission through stock assessment and a catch quota system, and later through total prohibition on certain species (in particular the Blue and Humpback Whales whose stocks have been reduced by ten- to fifty-fold or more). The history of whaling is a tragedy of excessive overfishing.

There are about 34 pinnipede species. Sea-lions and fur seals occur on coasts and islands of the North and South Pacific and of the southern Atlantic and Indian Oceans. Several of the species are associated with upwelling areas off desert or semi-desert coasts. The walrus is limited to the Arctic and feeds mainly on infaunal molluscs, but also takes crustaceans and echinoderms. The true seals are the most aquatic pinnipedes. Some, including the crab-eater seal, which in fact feeds mainly by straining krill between its teeth, are mentioned in Chapter 8.

Seaweed resources

The present commercial harvest of seaweeds is probably about 3×10^6 t fresh weight. This harvest is for human consumption, cattle meal and various industrial uses (Table 9.3). In Europe few seaweeds are eaten directly, but there is some demand for *Rhodymenia (Palmaria), Porphyra* and *Chondrus*, known locally and respectively as dulse, laver or sloke and carrageen. *Laurencia pinnatifida* is occasionally used as a condiment or chew, and biologists are known to eat the reproductive fronds of *Alaria*. More serious use of edible seaweed occurs in the Far East where nori (*Porphyra* or *Monostroma*), wakame (*Undaria*) and kombu (*Laminaria*) are traditional foods.

Seaweed has been harvested for fertilizer or soil dressing directly from the shore (in parts of Ireland boulders were laid out for this purpose) or from the driftline. It is still harvested for animal feed supplement in some areas of Canada and Norway. Formerly *Laminaria* and *Ascophyllum* were burnt to produce sodium-rich ash (*kelp*—a name later applied to *Laminaria*

Table 9.3 Current usage of seaweeds for the food and extractives industries. (From Jensen, in Jensen and Stein, 1979.)

Product (and algae used)	Value ($m y^{-1}$)	Weight produced (10^3 t. y^{-1})	Fresh weight harvested (10^3 t. y^{-1})
HUMAN FOOD			
Nori (*Porphyra*)	200	18	220
Wakame (*Undaria*)	200	7	60
Kombu (*Laminaria*)	200	100	700
ANIMAL FODDER			
Seaweed meal (*Ascophyllum*)	10	30	100
SEAWEED EXTRACTIVES			
Agar (*Gelidium, Gracilaria*)	40	6	150
Carrageenan (*Chondrus,* *Gigartina, Eucheuma*)	60	10	130
Alginates (*Macrocystis,* *Laminaria, Ascophyllum*)	65	15	400

plants). Kelp was exported before the mid-eighteenth century, for example from Portaferry to Bristol and Dublin, for use in glass manufacture and for the linen industry. However these and later uses, such as iodine extraction, have been superseded. Interest now centres on *phyco-colloid* production. These are polysaccharides, with emulsifying and gelling properties, derived from sugars or sugar-like acids. *Alginates* come from brown seaweeds and contain D-mannuronic and L-galactose polymers and *agars* are derived from red seaweeds such as *Gracilaria* and *Euchema*. Alginates are used for sizing paper and as a soluble yarn, but their main use is in food processing, for example in preventing large crystal formation in ice-cream. Amongst many other purposes agars (which contain D- and L-galactose and compounds derived from these) are used to texture food, to stabilize emulsions such as salad cream, to produce gels for electrophoresis and as a culture base for bacteria and fungi. Carrageenan, an agar-like compound from *Chondrus* (carrageen or 'Irish moss') is also used in clarifying beer. The possible sustainable world yield of seaweed is in the order of $15\text{-}20 \times 10^6$ t wet weight.

There is considerable interest in using seaweed for biomass production, for removal of excess nutrients from sewage treatment effluents and in multispecies aquaculture (see Jensen and Stein, 1979). Single-species culture of *Euchema* for industry is well established in Indonesia and the Philippines; culture of nori in Japan has been established for centuries. The most extensive culture is in China where *Laminaria japonica* nurseries have been established in the northern coastal province of Liaoning. Sporophytes are transplanted to other coastal areas where they can be grown during the winter months and harvested before the water becomes too warm (Cheng, 1968).

Environmental impact of fisheries

Relatively little is known about environmental effects of fisheries. Extensive removal of a species can lead to increased abundance of its competitors and its prey, for example the urchin–otter relationship, or the increase in crab-eater seals as euphausiid stocks rose due to overfishing of baleen whales. Many fishing methods are indiscriminate, thus there are large by-catches of non-target species. *Raja batis*, the common skate, is probably nearing extinction in parts of the Irish Sea since newly-hatched young are too large to pass through the permitted mesh of fishing nets. It is possible that other elasmobranchs are similarly threatened. Trawling can disrupt bottom communities, as in the case in *Modiolus* areas where dredging effects can persist for 12–20 years or more (Wiborg, 1946). The increasing use of suction or hydraulic jet dredges for burrowing molluscs such as *Mya* may destabilize bottom sediments. Bait digging can lead to local erosion of *Zostera*. Winkle picking is destructive when boulders are left overturned since both the upper (algal) and underside (epifaunal) communities are destroyed. Harvesting of *Ascophyllum* increases growth of the plants but can lead to considerable depletion of the associated fauna. Details of this and some other examples can be found in Kinne and Bulnheim (1980).

Human coastal communities benefit from fisheries in a number of ways. Clearly some villages owe their origin and continuance, if only at subsistence level, to the inshore harvest. Fishing villages often have good vernacular architecture and attract tourists. Fishing communities usually have a strong sense of identity derived, as in mining communities, from shared dangers, joys and concerns. Fishermen's wives are often involved in activities such as gutting, salting and selling fish. Fishing ports often attract associated enterprises—from the mid-1960s onwards as much or more of the world's fish catch has been sold processed than has been sold fresh, and marketing, carriage and retailing support further jobs, the number of which may exceed those in the fishing fleet.

CHAPTER TEN

COASTAL BIRDS

Anyone who has spent time by the sea cannot fail to have been impressed by the diversity and abundance of coastal birds. The vast flocks of waders, duck and geese that overwinter on estuarine mud-flats and the noisy crowded colonies of cliff-nesting sea-birds are particularly striking sights. This chapter briefly considers the major groups of coastal birds, the nature of their adaptations and their potential importance in coastal ecosystems.

Classification of coastal birds

Birds evolved on land but some soon turned to exploiting the resources of the sea, and it is amongst the marine representatives that many of the most primitive birds are to be found. Many species utilize the coast or the surface waters of the sea. For some the sea is their normal habitat and many of these *primary sea-birds* visit the coast only to breed. Although the sea covers some 70% of the earth's surface, only 3% (260–285 species) of all birds can be regarded as truly marine. Primary sea-birds belong to four major groups. These are:

(1) Order *Sphenisciformes* (penguins)
(2) Order *Procellariiformes*, or tubenoses (petrels, diving petrels, shearwaters and albatrosses)
(3) Order *Pelicaniformes* (pelicans, gannets and boobies, tropic birds, frigate birds, cormorants and darters)
(4) Order *Charadriiformes* (gulls, terns, skuas, skimmers and auks).

Secondary sea-birds are predominantly birds of fresh water and whilst they do resort seasonally to the coasts, generally to feed, they are not usually considered to be true sea-birds. They include divers, grebes, 'seaducks' (e.g. eiders, scoters, mergansers) and phalaropes. In addition to sea-birds, many waders (principally sandpipers, plovers and oystercatchers) and wildfowl (ducks, geese and swans) also occur within the coastal zone. Many of these, however, like many of the secondary sea-birds, temporarily forsake the coast for their inland breeding grounds. Several *land-birds* also utilize the coast and some of these (e.g. rock pipits) are rarely encountered elsewhere.

Coastal birds have also been variously classified according to where they feed and the ways in which they disperse (e.g. Dorst, 1974). '*Littoral*' *species* are restricted to the intertidal zone or to very shallow inshore water (except when migrating). Waders and most ducks and geese as well as the coastal land-birds belong to this category. *Neritic species* frequent the coastal waters overlying the continental shelf even during their migrations, which usually take the form of widescale dispersions. Examples include most gulls and terns, penguins, pelicans, cormorants and auks. *Oceanic species* are highly adapted to a truly marine existence. Whilst depending on land for reproduction, these are essentially birds of the open oceans, since even during the breeding season they make extensive oceanic feeding flights. Procellariform sea-birds belong to this category as do many frigate birds, tropic birds and gannets. Such ecological classifications are, however, somewhat artificial—even individual species (e.g. kittiwake) often span the boundaries of any system of groupings.

Distribution

Productivity throughout the world's seas exhibits marked variations according to climate and oceanographic conditions. In general, cold waters are more productive than tropical or subtropical waters, and inshore waters are more productive than the open oceans (see Chapter 2). The abundance and diversity of sea-birds reflect these global differences in production. Accordingly the world's seas can be broadly divided into distinct latitudinal zones largely on the basis of their characteristic avifaunas, though some species do transgress these zones during their migrations.

The austral zone surrounds the Antarctic continent and extends over the cold seas of the Southern Hemisphere. It is divided into two subzones by a distinct oceanographic boundary, the Antarctic convergence. Waters south of this line are colder and less saline than those to the north. The high productivity characteristic of these waters results in an especially rich avifauna in which penguins, albatrosses, petrels, shearwaters and, north of the convergence, cormorants predominate. Tropical and subtropical waters are less productive and their avifaunas correspondingly impoverished. In the tropical zone in particular, sea-birds often congregate over the continental shelf, thus avoiding the oceanic waters where production is particularly low. This avifauna between the tropics consists largely of terns, frigate birds, tropic birds, boobies, pelicans and, south of the equator, some albatrosses. Proceeding northwards through the cooler waters of the Northern Hemisphere conditions become progressively more favourable. The Arctic is comparable to the Antarctic in terms of production. Sea-birds of these colder northern waters, like those of the southern oceans, are therefore abundant and highly diversified. Amongst the more characteristic sea-birds of this zone are the gulls, auks (e.g. guillemots, puffins), skuas, petrels, cormorants, gannets and some terns. Whilst avifaunas from the Northern and Southern Hemispheres are not closely related systematically (though some species are common to both zones) they do exhibit some similarities due to convergent evolution (Figure 10.1).

Most northern sea-birds have markedly narrower ranges than their counterparts in the Antarctic. Fragmentation of the Arctic Ocean by continental land masses has encouraged the evolution of local races or even distinct species. The continuity of the southern oceans, on the other hand, has produced relatively homogeneous avifaunas with distributional zones which are essentially latitudinal.

The subdivision of the major oceans into broad latitudinal zones is modified by ocean currents and upwellings of cold water. Upwelling currents along the Atlantic coast of South Africa (Benguela) and along the

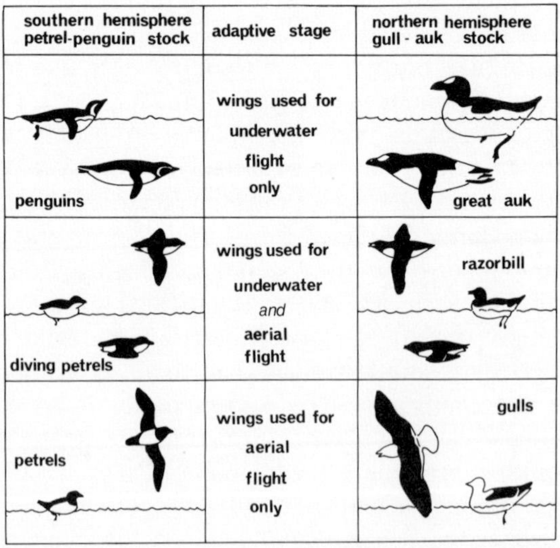

southern hemisphere petrel-penguin stock	adaptive stage	northern hemisphere gull - auk stock
penguins	wings used for underwater flight only	great auk
diving petrels	wings used for underwater and aerial flight	razorbill
petrels	wings used for aerial flight only	gulls

Figure 10.1 Convergent evolution in two stocks of sea-birds. (After Storer, 1971.)

Pacific coast of South America (Humboldt) carry rich Antarctic water far northwards. These cold currents are highly productive and enable especially dense avifaunas, often with many endemic species, to become established on adjacent coasts. The famous guano bird population off the Peruvian coast, for example, consisting mainly of cormorants, boobies and pelicans, contains several million birds (see p. 176). The intrusion of these cold currents into low latitudes also enables some typically Antarctic birds (e.g. penguins) to penetrate northwards even as far as the equator.

Upwelling areas and fronts (convergences) are often temporary and very localized. Consequently sea-birds, which flock to these areas of greatly enhanced productivity, are often very patchily distributed. Coastal waters too are especially productive and proportionately more sea-birds occur here than in the open oceans. The intertidal zone includes strongly differentiated habitats (Chapters 3, 4, 5) and supports diverse avifaunas whose composition depends largely on local ecological conditions.

Adaptations to the marine environment

All organisms are adapted to make maximum use of their particular environment. Birds are no exception, but marine species face some special difficulties not generally encountered by their terrestrial or freshwater counterparts.

Most sea-birds take in large amounts of salt in their highly specialized diets. Excess salt is excreted through *nasal glands* which lie in or over the eye orbits in shallow depressions in the skull. In microscopic structure these glands resemble kidneys. Their ducts discharge concentrated brine into the nasal cavities. In many species ridges along the edges of the beak direct this brine to the tip of the bill where it is unlikely to be swallowed. Nasal glands occur in all birds that frequent the sea or shore but their development largely parallels the extent to which they are committed to a truly marine existence and diet. They are thus more highly developed in oceanic than in 'littoral' species.

Sea-birds have evolved many adaptations to marine conditions including soaring, swimming and diving (e.g. Rosa, 1978). The long narrow wings of the albatrosses and shearwaters, for example, are modifications for prolonged periods of effortless flight. Auks, diving petrels and penguins, on the other hand, have wings that serve for underwater swimming and are thus largely unsuitable for economic flight. The wing bones of penguins are solid and flattened and some joints are fused, thus producing a very rigid structure. Many sea-birds dive, often to considerable depths (> 250 m in some penguins). These birds must overcome temporary shortages of oxygen. This is partially achieved by storing oxygen, both in an extensive system of air-sacs and in the large amounts of myoglobin present in the muscles. Many diving birds have a larger blood volume than their terrestrial relatives, further increasing the amount of oxygen that can be held in the body. Penguins also conserve oxygen by dramatically slowing their heart rate (bradycardia) during dives. At the same time some organ systems can be cut off from the blood supply leaving the available oxygen for those for which it is vital. Muscles use the oxygen stored in their myoglobin but when this is exhausted they can function anaerobically until the bird resurfaces. Many diving sea-birds can change the focal length of their lens, enabling them to see more effectively underwater.

Penguins, which spend much of their time fishing in extremely cold waters, are especially well insulated against heat loss. Their plumage is impervious to wetting whilst limb temperatures can fall dramatically without tissue damage. Compact scale-like plumage and a relatively high specific gravity help these diving birds to reduce their buoyancy. For the same reasons cormorants have wettable contour feathers and sometimes swallow pebbles before they dive; like many divers they also have large paddle-shaped feet set well back on their bodies. Plumage colour is also adaptive. The white belly and underwings of many sea-birds effectively camouflage them from their prey. White *frontal aspects* (foreheads and leading edges to the wings) are especially prominent amongst plunge-divers (p. 167).

An interesting adaptation for carrying food over long distances to developing chicks is found amongst the tubenoses. Partially digested food is stored as a concentrated foul-smelling oil. The weight of food carried is therefore greatly reduced and longer foraging excursions are possible. Stomach oil may also be used in defence or to waterproof feathers. Penguins secrete oil from a specialized part of the oesophagus.

Food acquisition

Coastal birds eat a variety of pelagic and benthic organisms and in many food chains they are top predators. Most pelagic sea-birds are fish-eaters, though squid and zooplankton may also feature prominently in the diets of some species (Nelson, 1980). Several feeding methods can be recognized (Figure 10.2). These include:

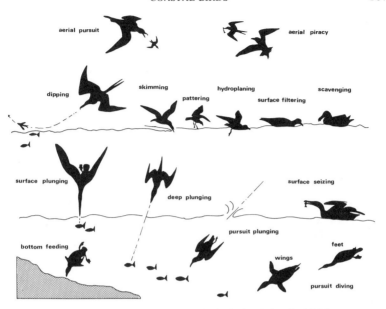

Figure 10.2 Sea-bird feeding methods. (After Ashmole, 1971.)

(1) Underwater swimming (using wings). Most species employing this method catch their prey by 'pursuit diving' in which the bird dives from the surface and pursues its prey underwater (e.g. diving petrels, penguins). Some species are 'pursuit plungers' (e.g. shearwaters). Here the bird plunges from the air but once underwater swims actively using wings and/or feet.

(2) Underwater swimming (using feet). Members of this group generally feed by 'pursuit diving' (e.g. cormorants, 'seaducks'). In shallow water some wing- and food-propelled divers feed on bottom invertebrates.

(3) Plunging (without swimming). This method is distinct from 'pursuit plunging' in that the prey is approached solely through the momentum gained before entering the water. Shallow plunging is characteristic of terns, pelicans and tropic birds, whilst deep plunging occurs principally in sulids (gannets and boobies).

(4) Feeding while settled on the surface. This category includes several distinct methods such as 'surface seizing' (e.g. albatrosses) and 'surface filtering' (e.g. fulmars). 'Hydroplaning', where the bird propels itself over the surface using its feet, is a special form of filtering exhibited by prions.

(5) Feeding at or near the surface whilst in flight. 'Dipping' (e.g. frigate birds, some terns) and 'skimming' (e.g. skimmers) fall into this category. Storm petrels feed on small zooplankton using a technique known as 'pattering'. Here the bird 'flutters' over the surface using both its feet and wings.

(6) Piracy, aerial pursuit and scavenging. Some species (e.g. frigate birds, skuas) force other birds to disgorge; others (e.g. skuas, large gulls) sometimes capture small birds in flight. Many gulls and petrels are scavengers.

The total range of feeding adaptations shown by sea-birds in any given locality effectively reduces interspecific competition and allows most of the available food to be exploited.

'Littoral' birds (p. 163), like pelagic sea-birds, are mainly opportunistic both in terms of food items taken and methods of capture. Some species are herbivores (e.g. kelp and Brent geese, widgeon) others, like egrets and herons, are fish-eaters, whilst flamingos are specialized filter-feeders. However, the vast majority of 'littoral' birds feed on benthic invertebrates. Waders, ducks and geese often occur in dense mixed flocks and competition probably occurs, especially when food is scarce—a situation which may be aggravated during cold weather when many invertebrates burrow more deeply and thus become largely inaccessible (e.g. Evans, 1979). Generally, however, competition is more apparent than real, since each species tends to specialize on particular groups of prey. Moreover, at low tide, birds are often distinctly zoned relative to the water's edge. Thus, on mud-flats, plovers haunt the drier sediments, dunlins the wet mud at the water's edge and redshank and godwit the shallow water. Food is therefore partitioned and many species are able to coexist.

Feeding methods and the type of prey taken are closely linked with bill morphology (Figure 10.3). Plovers and sandpipers have short bills and

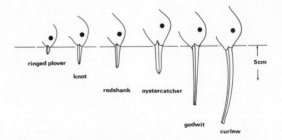

Figure 10.3   Bill length of some common European waders.

take mainly surface-dwelling invertebrates such as the mud-snail *Hydrobia* though larger prey such as worms are taken whenever these come to the surface. Dunlin, redshank and other waders with medium-length bills probe the surface layers of the sediments for worms, small crustaceans and bivalve molluscs. Only the longer-billed birds like curlew, willet and godwit can reach deeper prey. Oystercatchers have a particularly stiff but rather insensitive chisel-shaped bill ideal for hammering open bivalves. The diet and feeding behaviour of oystercatchers varies on different shores according to prey availability, differences in the strength of the prey and the firmness with which it is attached to the substratum. The bills of many wildfowl (e.g. shelduck), unlike those of waders which are adapted to take individual items of prey, have evolved to enable them to sieve large quantities of minute prey items. This technique, known as 'dabbling', clearly demands a high prey density (O'Connor, 1981).

Breeding ecology

All birds must come ashore to breed and are therefore arguably less well adapted than mammals to the marine environment. In this section we shall consider only the true sea-birds since, with few exceptions, most other coastal birds move to the Arctic tundra or inland during their breeding seasons. Reproductive strategies in birds have evolved to ensure that on average the greatest surviving number of young is produced. The main factors involved in this evolution are availability of food (for chicks and egg-laying females) and the risk of predation on eggs, young and adults. The effects of these factors may partly counteract and the strategy adopted is therefore often a compromise.

Virtually all sea-birds are colonial (98% compared with 13% for all birds) but two broad groups can be recognized (Lack, 1968). The first of these comprises all offshore- and some inshore- (i.e. within sight of land) feeding species. These breed in large colonies in relatively predator-free sites (e.g. cliff ledges, offshore islands, Antarctic mainland). Generally they do not need to defend their nests against predators and neither eggs nor chicks are cryptic. Penguins, tubenoses, pelicaniform species and auks belong to this group. The second group consists of the remaining inshore-feeding species. These nest in smaller, more accessible colonies (often on the ground) which they collectively defend. Their eggs and chicks (which usually leave the nest soon after hatching) are cryptic. This group includes gulls, terns, skimmers and skuas.

Whilst suitable nest sites in the case of the first group are generally safe from predators, they are often in limited supply. This leads to intense competition, though most species have evolved elaborate behaviour patterns to minimize the problems associated with group living. Intraspecific competition often forces younger inexperienced birds into marginal sites; some may even be prevented from breeding altogether. Interspecific competition results in the broad stratification of species which is commonly observed on many sea-cliffs, each species exhibiting a distinct preference for a particular type of nesting site, and in very early site selection in some species. This results in maximum site utilization. Nest sites are rarely limiting in the case of the second group and competition for them is therefore substantially reduced. Consequently these birds exhibit little segregation of nest sites. Their loose colonies result mainly from experienced birds returning to the same colony and to young birds joining already established colonies. Thus, whereas the first group, nesting in safe sites, is protected by inaccessibility, this second group relies on crypsis and cooperative defence to protect them from predators.

Whilst availability of suitable nest sites is probably the key factor favouring colonial breeding, feeding habits are also significant. Since most inshore feeders have widely dispersed nests within small colonies, each breeding pair does not need to go far to locate sufficient food. Thus some of the possible disadvantage of low-density nesting is outweighed by shorter feeding excursions. Offshore feeders *must* fly long distances to obtain suitable food so it makes little difference whether they nest in a few large groups or many small ones. By nesting in large colonies they can be especially selective in their choice of sites. Moreover, many of these oceanic birds are ill-adapted for life on land and they could therefore be particularly vulnerable to predators in all but the most inaccessible sites. Some albatrosses are ground-nesting but these are large birds well able to defend themselves and their chicks. Colonial breeding may also be important in 'information transfer'; larger numbers of birds could, for example, pool their knowledge concerning favourable feeding locations. Some sea-birds (e.g. skuas, gulls) will nest either solitarily or colonially depending on the availability of suitable food and nest sites. All sea-birds are normally monogamous and many retain the same mate in successive years. This allows pairs to breed earlier in the season and is probably the reason why it has evolved amongst so many sea-birds.

Many of the differences in breeding biology between inshore and offshore feeding families (Table 10.1) can be attributed to differences in the nature of their food supply. Offshore species travel further, often to a

Table 10.1. A comparison of the breeding biology of sea-birds (data mainly from Lack, 1968).

	Arctic and Common Tern	Sooty Tern	Black-headed Gull	Albatross[+]
Feeding habitat	inshore	offshore	shore/inshore	offshore
Usual clutch	2(3)	1	3	1
Incubation period (days)	21	$29\frac{1}{2}$	23	64–79
Fledging period (days)	30	60	30	140–280
Incubation spells	$\frac{1}{3}$–2 h	$5\frac{1}{2}$ days[++]	1-$2\frac{1}{2}$ h	6–21 days
Feeding visits to young	3–4 per h	1 per day	1 per h	1 every 2–3 days
Age at first breading (y)	2–4	6	2	5–11
Annual adult mortality (%)	25	16–20	18	c.3

[+] Based on data for several species of *Diomedea*.
[++] Based on 2 unusual years; probably closer to 1 day.

sparser and/or less predictable source than do inshore feeders. Consequently offshore birds produce smaller clutches (usually one egg), visit the nest less often and have longer incubation and fledging periods. However, their large eggs (which are often tapered so that they do not roll off cliff ledges) hatch into chicks at an advanced stage of development. These downy chicks are well protected against the cold and their large food reserves enable them to withstand prolonged periods without food. Offshore-feeding birds also generally start to breed at a later age than inshore birds, again possibly reflecting the greater difficulties these species experience in raising young. Their lower reproductive rates, however, are offset by lower annual adult mortality (Table 10.1). Inshore-feeding birds by contrast operate in an environment where food is more abundant and more predictable but where predation can be intense. Accordingly they produce bigger clutches (2–3 eggs) and raise chicks with shorter incubation and fledging periods.

In sea-birds living in highly seasonal environments breeding is generally timed to coincide with the maximum standing crop of their prey. In the tropics, where food supply may exhibit little or no seasonal variation, breeding is frequently non-annual, irregular or even continuous. Even so, reproduction may still be highly synchronized (possibly via social stimulation) so that all chicks within the colony are at a similar stage of development. By synchronous breeding the survival chance of any individual chick is substantially increased, since predators are effectively 'swamped' with large numbers of prey items of comparable profitability.

Mortality, longevity and regulation of numbers

The low reproductive rate of many oceanic birds is accompanied by high adult survival and a prolonged reproductive life. Reliable estimates of survival are understandably difficult to obtain but evidence strongly suggests that many of these birds are exceptionally long-lived. Annual mortality rates of 3–5% are not uncommon and potential life spans are probably well over 50 years (a 53-year-old Laysan Albatross has been recorded). Heaviest mortalities occur amongst younger, less experienced birds, but once they are beyond their first year survival is not markedly age-specific. Mortality is often heavier in male birds, presumably reflecting their involvement in territorial disputes. Starvation, often exacerbated by prolonged spells of bad weather, is the major cause of death. Disease, parasites, accidents and predation are less important sources of sea-bird mortality.

Estimates of wader mortality based on returns to nest (7.5–30.0%) are approximately half those based on ring returns (Hale, 1980). Evidence suggests that most adult deaths probably occur during winter as a result of cold weather and starvation and that smaller species such as dunlin, redshank and knot are especially vulnerable (O'Connor, 1981). Juvenile birds often appear to be most at risk, probably because they feed less efficiently or because they are forced by adults into suboptimal habitats.

In general, bird populations are regulated within certain limits by density-dependent factors. These factors are thought to affect recruitment to the breeding stock and death rate, usually the latter of these, since clutch size is often defined. Recruitment can, however, be affected if at high populations less experienced birds are prevented from breeding. It has been argued that most, if not all, density-dependent mortality probably occurs during the breeding season and chicks and juveniles are most likely to be involved. The principal limiting resource is probably food. When breeding, birds must constantly return to their nests. They can therefore exploit only that area around the nest site which they can search economically. Although this varies from species to species it represents only a fraction of their potential food stocks. As the breeding population increases in size, competition may heighten to a point where adult birds are unable to obtain sufficient food for themselves and their young. Breeding success then declines to a point where the remaining population can be sustained. Population increase in thus halted and may even be reversed if food becomes further restricted. Once breeding is completed, however, birds are free to disperse over a wider area where food is less likely to be limiting.

Factors other than food may also effectively regulate population density. Predation is probably of minor importance for most coastal birds, though avian predators (e.g. skuas, gulls) may occasionally take a heavy toll of eggs and young. Ground-nesting species are especially sensitive to mammalian predators (e.g. foxes, rats) and there is evidence that egg and nest losses in some waders may be heavier at high population density. Nest sites may also be limiting, and large population increases often occur when birds gain access to new breeding areas. Some sea-birds may never be food-limited, so vast are their potential feeding grounds. This has led to the suggestion that some form of social regulation may effectively keep populations broadly in balance with available resources.

Migration and foraging

Some birds migrate to make the best use of a changing environment. They move from areas which are likely to become difficult to more favourable areas in which their survival is substantially enhanced. Very few birds are tied to a single locality; even penguins, which are flightless, travel considerable distances in response to fluctuating environmental conditions. The pressure to migrate is greater in highly seasonal environments. Migratory patterns are complex and vary within and between species. For example, within any given species there may be migratory and non-migratory populations and the latter may undertake journeys of vastly different distances.

Although some birds remain faithful to the coast throughout the year, the appearance of most species in coastal waters is highly seasonal. Two broad trends can be discerned. (1) True sea-birds come ashore to breed, often seeking out remote nesting sites. (2) Most waders, duck, geese and secondary sea-birds use the coast seasonally, mainly to exploit the abundance of food, especially during the winter months when climatic conditions on their breeding grounds may be inhospitable. Some of these species do breed close to the shore (e.g. oystercatchers, ringed plovers, shelduck) or by brackish waters (e.g. avocets, flamingos) but the vast majority move inland, often moving northwards along traditional routes to their breeding grounds in the tundras of the Arctic (e.g. knot, sanderling, turnstone, Brent goose). This change of habitat involves switching diets and many waders which feed on marine invertebrates in their wintering quarters are predominantly insectivores when breeding. Migration, either to or from breeding sites, therefore spreads the population over a wider geographical area, thus allowing the exploitation of resources which would otherwise remain inaccessible. Several types of movement can be recognized, all of which are largely associated with locating adequate food supplies. Although these categories are convenient they are far from rigid.

(1) *True migrations* are predictable seasonal movements between breeding and wintering locations. The distance and direction of these movements are largely or wholly innate. Some birds undertake prodigious migrations, though the precise reasons for so doing still remain obscure. The trans-equatorial migrations of the Arctic tern (the most northerly populations of which nest in the high Arctic and winter in the austral summer as far south as the Antarctic ice floes) and the transoceanic migrations of many petrels and albatrosses are especially impressive journeys. The short-tailed shearwater follows a complex migratory route (Figure 10.4) overwintering in the North Pacific (actually the Arctic summer) and along the shores of the Arctic Ocean before returning to south-eastern Australia and Tasmania to breed. During this migration these birds make full use of the prevailing winds. They pass rapidly through the tropics where food is scarce. Manx shearwaters show the reverse pattern, breeding in the North Atlantic and overwintering off the southeast coast of South America. Many Arctic waders winter as far south as South Africa (Figure 10.4).

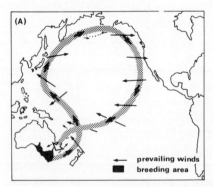
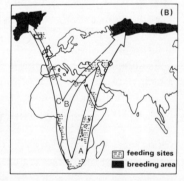

Figure 10.4 Migratory routes of (*A*) the short-tailed shearwater (after Nelson, 1980); (*B*) curlew sandpiper and sanderling. *A*, northward and southward routes of sandpiper and southward route of sanderling; *B*; northward route of sanderling; *C* sanderling route to Greenland. (After Branch and Branch, 1983.)

(2) *Dispersal movements* are less directional than true migrations. Outside the breeding season many sea-birds have no specific wintering areas. Instead they move in a more or less dispersed manner, often congregating and forming temporary bases wherever food is locally abundant, for example at convergences. Tropical birds in particular disperse widely, although often without any strong directional component to their movements. Dispersal is also characteristic of many auks, penguins and cormorants; in some gulls it occurs on a very limited scale. *Eruptions* (mass emigrations to areas not normally visited) are similar to dispersal movements. They are generally weather-induced and are not regular phenomena.

(3) *Nomadism* differs from dispersal in that birds wander more or less continuously, covering large distances, often along broadly defined routes, that eventually return them to their breeding sites. Nomadic wandering is essentially a response to the vast areas over which many oceanic sea-birds (e.g. tubenoses) have to search for their food.

(4) *Foraging movements within the breeding season* can vary in distance and can be either strongly directional or more random. Some breeding sea-birds (e.g. shearwaters, petrels) forage far (up to 1000 km) from their breeding colonies. Gannets and many auks (such as guillemots) range widely over the continental shelf, whilst many tropical birds remain at sea for many days. The distance over which these birds have to fly in order to obtain appropriate food for their developing chicks is a major factor determining breeding strategy (p. 170).

The importance of birds in coastal ecosystems

Coastal birds often occur at exceedingly high densities, and since the vast majority are predators their potential impact on the ecology of the intertidal zone and inshore waters is considerable. Detailed quantitative studies of this impact, however, are surprisingly scarce. Estuarine birds, particularly waders and duck, eat large numbers of invertebrates. Individual redshank can consume 40×10^3 *Corophium*, oystercatchers over 300 cockles, and knot over 700 *Macoma* each day, and flocks of these birds frequently number several thousand. In the Dutch Waddenzee the eider population alone reportedly takes over 1×10^9 cockles, 2.2×10^8 mussels and 30×10^6 crabs annually; oystercatchers account for 3×10^9 cockles and sandpipers over 1×10^9 worms. Flocks of shelduck can eat 15×10^6 *Hydrobia* during each feeding cycle. Clearly the quantity of food required to sustain such populations only occurs in extremely productive environments.

The effect of such depredations on the prey population is variable. For some species it may be negligible whilst for others it may be more significant. Oystercatchers regularly remove around 40% of the total cockle production (considerably more for some size classes) during the course of the winter. In the Ythan estuary in Scotland almost the entire annual crop of mussel flesh was consumed by winter (Milne and Dunnet, 1972), some 37% of this going to birds (mainly eider, gulls and oystercatchers). Such figures emphasize more clearly than absolute numbers the major role of shore-feeding birds as consumers of invertebrate biomass.

In general, rocky shores do not support the large flocks of birds seen on estuarine mudflats (Feare and Summers, 1985). Large numbers of small snails are taken by rock pipits and purple sandpipers; the latter have been estimated to be responsible for 93% of the 89% winter mortality of first-year dogwhelks (*Nucella lapillus*). Turnstones feed on barnacles and, like many other coastal birds, on the rich harvest of insects and amphipods associated with the banks of rotting seaweed on the high-water mark. Oystercatchers take limpets and mussels. Limpets living in rock pools are often more heavily predated than those which are tightly clamped down on the exposed rock surface, thus emphasizing the importance of prey microhabitat. Some limpets cease to be vulnerable to predation once they have attained a size beyond the bird's capabilities (i.e. 'refuge in large size').

Many shore-birds exhibit distinct preferences for certain prey species, often depending on their local abundance (e.g. Schneider, 1978). Even within any given prey species particular size ranges are frequently favoured according to their profitability. Consequently birds can have a marked effect not only on the relative composition of benthic communities but also on the population structure of individual prey species.

The effect of birds on pelagic communities is more difficult to ascertain. In the immediate vicinity of their breeding areas they may temporarily over-exploit their prey. Outside their breeding areas, however, it seems unlikely that sea-birds will have much predatory impact, so vast and productive are their potential feeding grounds. Sea-birds and man may compete directly (see Furness, 1982). At the peak of the Peruvian anchoveta fishery, around 7% of the estimated sustainable yield was lost to guano birds. However, the combined effects of overfishing and El Niño (incursions of tropical water) in the early 1970s caused a massive reduction in both anchovies and guano birds. Sea-birds off the Oregon coast account for 22% of the annual production of small, pelagic, inshore fish; Shetland sea-birds take around 29% of the fish production within a 45 km radius of their colonies. Sea-birds are also known to be important consumers of herring in the east Bering Sea.

Birds therefore remove large quantities of food from the coastal zone, though some of the contained nutrients are recycled via faeces. In regions where these are deposited in quantity they are sometimes mined commercially as guano, one of the purest sources of phosphate.

CHAPTER ELEVEN

COASTAL MANAGEMENT

Most of the previous chapters have considered the ecology of the coastal zone as if man were merely an observer. This is rarely true. Man's abundance and activities have many ecological effects. These may be localized or widespread, temporary or permanent. A high proportion of mankind lives in the maritime zone and many major cities are on the coast. This arises because nearly all man's biological and social needs are readily provided by coastal-zone resources. It has been estimated that by AD 2000 75% of the population of the United States will be living in the coastal zone. The need for management arises because resources are finite. Some, such as many mineral and fuel resources, are *non-renewable*; others, such as plant and fish stocks, may be *renewable* but limited to particular areas. The need also arises because some of man's activities prejudice the continuing use of resources. Many of these detrimental activities differ from overfishing because they result from additions to, rather than removals from, the environment. Damaging added-substances are usually called *pollutants*.

A list of intertidal and inshore activities is presented in Table 11.1. These could be reclassified into four main types of resource provision, namely physical, mineral, organic (Chapter 9) and spiritual; the latter includes coastal zone recreation, tourism, education and aesthetics.

Physical resources

Physical resources of the coastal zone include the climate, which is often more agreeable to man than that of more inland areas. Apart from providing a suitable physical habitat, perhaps the oldest use of coastal physical resources is that of the sea for transport. Even in prehistoric times there were extensive waterborne migrations of peoples and their cultures along coasts and from island to island. Pre-classical examples are the Minoans, the Mycenaeans and the Phoenicians. Shortage of terrestrial

178

Table 11.1 Coastal activities. A classification of man's use of the shore and adjacent land and waters (largely from Shaw, in Anon., 1983).

Harvesting naturally occurring living resources

Mammals

Seals

Fish

Seine netting (e.g. plaice)
Purse seining (e.g. herring, mackerel)
Long line fishing (e.g. halibut)
Trawling for demersal fish (e.g. plaice, cod, skate)
Drifting (e.g. salmon)
Trapping with fixed devices or nets (e.g. salmon, eels, flounder)

Crustaceans

Lobster and crab potting
Scuba diving (crawfish and lobster)
Tangle-netting for lobster and crawfish
Trawling for shrimps, Norway lobster and prawns

Molluscs

Hand gathering (e.g. mussels, ormers)
Dredging (scallops and oysters)
Digging (e.g. clams, cockles)
Hydraulic dredging (cockles)
Trawling (queen scallops, squid)

Other invertebrates

Worms for bait
Sea fans
Sea urchins

Birds

Wildfowling
Egg collecting

Algae

Gathering from the littoral zone
Cutting from the sub-littoral zone
Gathering algae washed ashore
Consumption by sheep (e.g. in Shetlands)

Other Plants

Reeds for roofing
Plants for food (cattle, sheep and horse grazing on marshes)

Cultivation of living resources

Fish

Trout	Eels
Turbot	Salmon
Lemon sole	Sole
Brill	Plaice
	Cod

Crustaceans

Lobster

Molluscs

Oysters	Mussels
Clams	Queen scallops

Algae

For biomass	For chemicals
For food	(e.g. alginates, agar)

Farming and forestry

Extraction of non-living resources

Sand and gravel

From the sub-littoral
From beaches

Calcareous material

Shell sand

Water and its contents

Desalination
Extraction of chemicals

Energy (excluding oil and gas)

Wind energy

Table 11.1 (*Continued*)

Maerl	Tidal energy
Metalliferous sediments	Wave energy
Oil and gas	*Marine archaeology*
Coastal mines	

Waste disposal

Sewage	*Thermal discharge*
Domestic refuse	*Dredging spoil*
Refuse and oil from ships	*Radioactive waste*
Biodegradable industrial waste	Strontium
Inorganic and persistent organic industrial waste	Plutonium
	Caesium
Colliery waste	
China clay waste	*Incineration at sea*
Heavy metals	
Fly ash	
Acid, alkalis, phenols etc.	
PCBs, DDT etc.	

Use of coastal land

Coastal industrialization and urbanization	*Recreation*
Manufacturing industry	Holiday camps
Ship and oil-rig construction	Car parks
Housing	Rock climbing
Dockside facilities, warehouses, container parks	Sand yachting
	Camping and caravanning
	Collecting
	Bird watching
	Access (by car or pedestrian)
	Military activities on land
	Firing ranges

Use of water space

Enclosure of estuaries	*Recreation*
For fish farming	Sailing
For docks and marinas	Bathing
For freshwater storage	Scuba diving and snorkelling
By causeways for roads or railways	Sea angling
	Power boating
Infill	Surfing
	Collecting
For agriculture	Bird watching
For airports	
For port installations	*Transportation*
For industry	
For housing	Spillage from ships
	Collisions

Table 11.1 (*Continued*)

Artificial island construction	Refuse disposal
For industry	Tank cleaning
For airports	Dredging to improve navigation
For ports	
For waste disposal	*Military activities affecting water space*

Miscellaneous

Introduction of alien species	*Coastal defence and maintenance*
For coastal stabilization (*Spartina* grass)	Breakwaters and seawalls
By accident (e.g. slipper limpet, oyster	Stabilization of river mouths
tingle, Japanese seaweed).	Dredging
For cultivation (e.g. Pacific oyster).	Groynes
	Stabilization of shingle banks
Reintroduction of species	
	Scientific studies
Control of predatory species	
Seals	Collecting samples
Salmon-eating birds	Incidental damage (e.g. access by vehicles)
Shellfish-eating birds	Experimental manipulation
Natural processes	*Education*
Erosion	Collecting samples
Siltation of estuaries	Incidental damage (e.g. trampling, noise)
Waves	Experimental manipulation
Tides	
Accretion	
Storms	
Currents	
Immigration of new species	

resources drove the ancient Greeks to develop seafaring, which helped transform them 'from primitive tribes into bearers of a culture which was to dominate the then known world' (Van der Heyden and Scullard, 1959). Arabs, Chinese, Polynesians and Scandinavians also became great seafarers. The linking or, sadly, the destruction of peoples and cultures moved into a new phase with the transoceanic voyages of the late fifteenth and early sixteenth centuries. Coastal sea trade has declined in importance since the invention of the internal combustion engine, but transoceanic shipping remains of great importance in world trade, with about 9.5×10^6 t of goods and materials being shipped per day.

Space

Many coastal villages, albeit founded for fishing, trading or strategic reasons, have become major settlements demanding further resources including space. When land becomes limiting there is a demand for 'reclamation' of marginal, intertidal or subtidal areas for housing, industry and agriculture. *Reclamation* is a term to be deplored in the context of conservation. It implies that something previously lost is being regained, whereas the practice can lead to a net loss rather than gain of environmental resources. *Infill* is a term with less pre-emptive judgement of the issues involved.

In the United States about 7% of the estuarine and associated wetland habitat has been filled in or dredged out; this figure masks the intensive change which has occurred in some areas, for example the State of California has an estimated 67% loss. In Bermuda about a third of the then existing mangrove swamps were filled in for Air Force base construction in the early 1940s, and associated dredging led to the decline of several coral reefs (Figure 11.1). Similarly in the British Isles over one-third of the original foreshore of Belfast Lough has been filled in, and in the Tees Estuary less than 1% of the original tidal flats remain. The most famous 'reclamation' works are those of the Dutch (*Deus mare Batavus littora fecit*). About 18% of the present land area of the Netherlands was once intertidal or subtidal; two of the largest schemes were the infill of the Zuiderzee and the impoundment of the Rhine–Maas estuary by the Delta Works. The latter scheme was partly modified on conservation grounds, and it is likely that much of the Waddenzee area will remain open because of its importance as a fish nursery ground and its ornithological interest

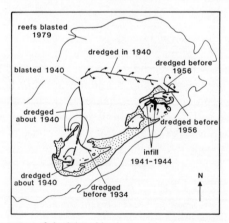

Figure 11.1 Major events of dredging, blasting and land reclamation around Bermuda. Arrows show where dredged material was deposited. Sediments and cloudy water caused by these activities stunt the growth of Bermuda's coral reefs. (After Sterrer and Wingate, in Hayward *et al.*, 1981.)

(see Zijlstra, 1972). Many coastal industries are the same as those further inland but are sited on the coast because of factors such as better access to transport. Some, such as shipbuilding and bromine or magnesium extraction plants, are sea-dependent. Several industries such as electricity generation require a large cooling-water supply and this is most readily met from the lower reaches of rivers or from the sea. The thermal effluent is rapidly dissipated within the much greater volume of the receiving water. Small populations of exotic warm-water species have become established near some of these effluents in temperate countries.

Power generation

The physical attributes of the sea can be used to generate rather than dissipate energy. In the 1970s various wave-energy devices including the Salter 'nodding-duck' and the Wells turbine were developed. It is thought that arrays of such devices would be particularly useful for power

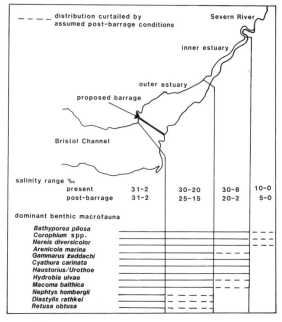

Figure 11.2   Predicted effects of a barrage for power generation on some major benthic invertebrates of the Severn Estuary, Britain. (After Mitchell and Probert, 1981.)

generation for remote coastal communities. Considerable attention has also been given to tidal electrical generation. The scheme at La Rance in northern France has been functioning since the late 1960s and contributes about 470×10^6 kWh y^{-1} to the French National Grid (Cotillon, 1978). The scheme was undertaken before the advent of environmental impact studies, but the proposed Severn Barrage in the upper reaches of the Bristol Channel, England has engendered considerable study including computer modelling on the GEMBASE system (Figure 11.8). The barrage could generate about 6% of the United Kingdom's electricity and is attractive since it uses a renewable energy resource, although it would cause some local and more extensive ecological changes (see Shaw, 1980). Some of these, resulting from altered salinity, are shown in Figure 11.2. At present the very high construction costs (approaching £6000 m) and the lengthy construction period (15y) militate against the proposal, but major planning decisions are not necessarily taken on economic criteria.

Mineral resources

Maritime and inshore areas are sometimes rich in mineral resources including sand, gravel and lime-rich deposits such as limestone or maerl (derived from the calcareous marine alga *Lithothamnion*). Extraction of minerals from intertidal or subtidal areas can cause coastal protection problems, for example by starving dune areas of their sand supply, or may lead to turbidity and environmental decline. Coastal mining gives rise to spoil heaps, and coastal extraction of chemicals such as salt, bromine, magnesium or water from the sea produces industrial effluents. The offshore extraction of oil and gas from areas such as the North Sea and the Gulf of Mexico produces various additional interactions between man and the coastal environment.

Pollution

The concentration of people and industry in the maritime zone inevitably leads to problems of waste disposal. The average adult in a developed country produces about 300 g of faeces per day. In the United States the average household produces 100 l of sewage per day but the average amount of garbage is 5 kg per person. In a year a town of 5000 people produces enough garbage to be impacted to a depth of 1 m over 1 hectare. It is hardly surprising that many coastal wetlands have been buried by refuse disposal. Other particulate waste causes various problems such as smothering of biota by increased sedimentation. Colliery waste tipped on beaches in County Durham, England, effectively prevents recreational use of about 9 km of shoreline and harms the local inshore fishery (Countryside Commission, 1970). However, such effects are small compared with the impact of refuse tipping and with the chemical effects which can follow from the discharge of particulate and dissolved organic or inorganic matter. At present the most dangerous chemical pollutants appear to be pesticides (and related organic compounds) and heavy metals. Other factors causing concern are radioactivity, oil, organic matter with high biological oxygen demands and excess nutrient supply. Fuller accounts can be found in Gerlach (1981) and Kinne (1982–3)

Pesticides and related compounds

Many of the wide range of pesticides in regular use are *organo-chlorine compounds*; related compounds are also used in plastic and other industrial processes. These chemicals are often referred to by their initials, such as PCBs (polychlorinated biphenyls), HCH (hexachlorocyclohexane) and DDT (dichlordiphenyltrichlorethane). Problems arise because these compounds are extremely toxic. They reach the sea via fresh water from run-off, via effluents, or occasionally via the atmosphere. Their concentration in inshore waters, especially estuaries, is often 5–10 times

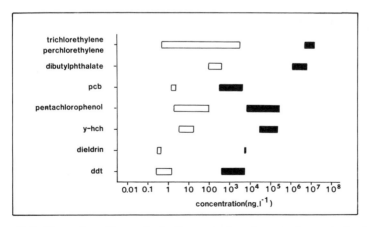

Figure 11.3 Comparison of known levels of concentration of some toxic organic chemicals in sea water (unshaded) with levels known to produce lethal effects in laboratory experiments (shaded). (From Ernst, in Kinne and Bulnheim, 1980.)

greater than offshore. This is not only because of their discharge routes but also because of the smaller dilution factor in the enclosed estuaries and bays where towns and industries are often sited. However, it is rare for *acute* toxicity levels to be reached in the receiving water itself because dilution factors are still considerable (Figure 11.3). Values for toxicity are usually measured as LC_{50}, that is the concentration of a substance which will kill 50% of the test animals within a given period (normally 48 or 96 h). However, in their natural environment organisms may be subject to pollutants for considerably longer periods of time and suffer from chronic rather than acute toxicity. *Chronic* effects (that is malfunction rather than death) are often shown at concentrations 10–100 times less than those producing acute response.

Two other factors add considerably to the danger of pollution from pesticides and heavy metals. They are *bioconcentration* and *biomagnification*. The first describes the retention of a pollutant in body tissue so that its concentration is increased above the environmental level; it comes about through absorption and other body processes. The concentration factors are often between 10^3 and 10^4. Biomagnification refers to the process of concentration through successive levels of a food chain or web. The degree of increase say from primary producer to primary consumer will depend both on the latter's feeding and metabolic rates. Magnification factors are not well understood but seem to range from tens to single figures between each trophic level; much of the evidence for this is circumstantial—predators often containing higher concentrations than prey (Bryans, in Kinne, 1982–4).

Heavy metals

Unlike pesticides, some heavy metals are essential requirements of life—for example haemocyanin, the blood pigment of crustaceans, contains copper. Marine organisms are adapted to obtain their essential heavy metal requirements from the low levels naturally present in sea water. Pollution problems arise because man's activity can increase these levels to toxic concentrations which interfere with essential biological processes such as enzyme function. The metals causing most concern in aquatic environments are Hg, Cu, Cd, Zn, Pb and Cr. Mercury was the first of these to be brought to worldwide attention. Most mercury compounds decompose in the sea to give mercury, mercuric chloride or mercuric sulphide, but some of these are converted to methyl mercury which is extremely toxic. In humans it affects the nervous system, causing impaired vision, hearing and speech and loss of muscular coordination. These symptoms appeared with tragic results in fishing communities around Minimata Bay, Japan, in 1953 and were epidemic by 1956. Fifty or more people died. The mercury had entered the sea from an acetaldehyde plant, where it was used as a catalyst, and had been bio-concentrated and -magnified through the food chain leading to fish and man. The safe human ingestion level is about 0.03 mg Hg day^{-1}; Minimata disease occurs at levels above 0.3 mg. Man produces about 9000 t Hg y^{-1} of which 5000 t enters the sea, mainly from chemical, paper and agricultural (seed dressing) industries. The average concentration in sea water is about 0.03 μg l^{-1}; polluting levels are about 0.2 μg l^{-1}.

There is no clear evidence that disease has been caused by any other seafood-derived heavy metals. A few instances are known where high levels of Cd, Cu and Zn in oysters have caused nausea and vomiting in humans. Copper enters sea water via mining and chemical wastes and is a common ingredient of antifouling paints. Cd, Cr and Zn uses include plating and galvanizing other metals. Cadmium is also used in the plastics industry. Lead has a large number of uses, one of which has been as an automobile fuel anti-knock ingredient, tetra-ethyl lead bromide. Lead levels in the sea are now about five times greater than in prehistory. About 1% of the world's lead production enters the sea each year.

Anthropogenic sources of heavy metals include mining and subsequent processes such as smelting, coal burning, ash disposal, metallic corrosion and antifouling paints. The latter represent a direct route of entry to the sea and contribute to high pollution levels in docks and harbours and, subsequently, in areas where sediment dredged from these is dumped. Other routes of entry are via freshwater input and also through the atmosphere. The latter is via gas, vapour or very small light solid or liquid (*aerosol*) particles (see Kullenberg, 1982). Atmospheric input may be considerable, thus the global annual input to the sea of about 300 000 t Pb includes 40 000 t via this route.

Radioactivity

There is a natural background level of radiation, but the concentration of radioactive substances in the sea is being increased by man. Initially this was due to a few large pulses from nuclear explosion fall-out, but the increase is now mainly due to nuclear power plants. Some also enters from

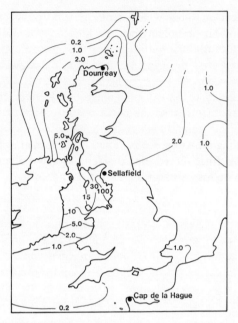

Figure 11.4 The distribution of caesium-137 (pCi l^{-1}) around the British Isles in 1973. The major source is Sellafield (Windscale) and from here material is transported mainly northwards around Scotland and into the North Sea. (After Anon, 1981.)

radioisotope use and from nuclear-propelled vessels. The pollution problem arises because radiation damages reproductive and somatic cells, thus giving rise to chromosome aberrations which may lead to cancer or other malfunctions. The dangers which arise from radioactivity have lead to the adoption of national and international standards for measurement and control. However, the safety regulations have not always been enforced and there has been great public concern both over licensed and accidental discharges, for example, those into the Irish Sea from Sellafield (Windscale). Radioactivity from this source has been used to trace water and sediment movements (Figure 11.4). A good account of marine radioactivity is given by Woodhead (in Kinne, 1982–4).

Oil

Coastal pollution by oil has received far more attention than its relative damage merits. This is probably due to the spectacular nature of oil tanker wrecks, the effects on birds which evoke a wide public response, the nuisance of oil and tar on holiday beaches and the money available from oil-based companies for appropriate research. Ideally oil spills should be contained and removed mechanically at source. However, if this fails, unless large populations of coastal birds or tourists are likely to be affected or fisheries are likely to be tainted, oil spills are best left to degrade naturally. Other problems arising in relation to oil and gas include the onset of anaerobic benthic conditions around sea platforms due to bacterial growth, and drilling mud deposition. This causes faunal impoverishment and increases platform corrosion problems. Offshore structures and resulting bottom debris can also interfere with fishing.

Organic matter

Organic matter entering water can cause environmental 'stress' through sedimentation, nutrient addition or through its *biological oxygen demand* (BOD). The latter is a measure of the use of dissolved oxygen by micro-organisms as they decompose the matter. BOD is used widely in water-quality studies and is usually measured by incubating samples at 20°C for 5 days and expressing the oxygen consumption in parts per million (BOD_5). In the United Kingdom the Royal Commission on Sewage Disposal has set standards for discharge of not more than 30 ppm for suspended matter and

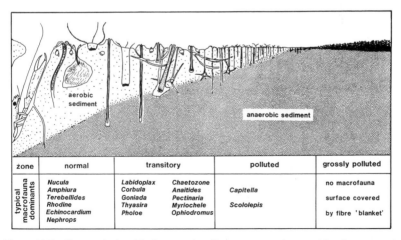

zone	normal	transitory		polluted	grossly polluted
typical macrofauna dominants	Nucula Amphiura Terebellides Rhodine Echinocardium Nephrops	Labidoplax Corbula Goniada Thyasira Pholoe	Chaetozone Anaitides Pectinaria Myriochele Ophiodromus	Capitella Scololepis	no macrofauna surface covered by fibre 'blanket'

Figure 11.5 Changes in benthic fauna and sediment oxygenation resulting from organic enrichment by paper mill pulp fibre. (From Pearson and Rosenberg, 1978.)

20 ppm BOD_5 at 65°F (18.3°C). However experience has shown that if receiving fresh waters are to be treated for subsequent human re-use a BOD_5 below 4 ppm is desirable.

The effects of large quantities of organic matter on aquatic communities include the decline of species diversity (see later), an increase in the relative abundance of deposit feeders, an increase in the number of smaller organisms and decrease of the larger organisms present and, unless the pollution is really severe, dominance of the macrofauna by a few or a single annelid species. In fresh water *Tubifex tubifex* is often abundant. This oligochaete has a tail well supplied with blood vessels and can extend it above the sediment towards oxygen sources. A similar habit is found in the corresponding marine polychaete *Capitella capitata*, which, together with small spionids, is also characteristic of organically enriched sediments (Figure 11.5).

Nutrients

The standard method of treating sewage usually includes removal and mechanical breakdown of larger particles (primary), oxidation of organic material (secondary) and sterilization of effluent (tertiary treatment). This will remove about 90% of the organic matter and bacteria as 'activated sludge' which is then disposed of on land or at sea. However dissolved nutrient levels in the effluent remain very high unless there is further biological or chemical treatment. In fresh water, estuaries and enclosed sea areas, increased nutrient levels may enable plankton blooms (Chapter 2) to continue throughout warm weather, thus upsetting natural cycles and leading to eutrophication. Some green seaweeds such as *Enteromorpha* can become superabundant in shallow areas receiving sewage effluent. The death and decay of phytoplankton blooms and of such seaweeds then lead to water deoxygenation. Anaerobic conditions lead to release of sediment phosphate (Chapter 4) and, if this has been a limiting nutrient, a positive feedback is established.

Public health

In some bloom conditions one or more species dominate the marine phytoplankton. Unfortunately some of the forms which can do this, such as the dinoflagellates *Gonyaulax* and *Gymnodinium*, produce neurotoxins and can produce 'red-tides' causing extensive mortality of other species including fish; they can also lead to paralytic poisoning of people who eat shellfish that have fed during the bloom. Another possible health hazard from sewage is infection by gastro-enteritis or paratyphoid-causing agents. The Eighth European Marine Biological Symposium scheduled for Naples in 1973, for example, had to be cancelled due to a typhoid outbreak. However, the chances of gastric infection from consuming fish or shellfish or indeed from swallowing sea water when swimming are low except in grossly polluted areas. Bathing beach designation in the EEC countries requires that there are less than 2000 coliform bacteria per 100 ml of water and that in general the standard should be less than 100 coliforms in 100 ml. Fortunately, most enteric organisms are short-lived in sea water, especially in warm and light conditions; unfortunately, some can survive once they are inside shellfish and can be concentrated there to levels ten times above the environmental density.

Pests and introductions

Several marine and estuarine species, such as the Australian barnacle *Elminus modestus*, or the Chinese mitten crab *Eriocheir sinensis*, have become established in areas far from their original home. This has happened either accidentally or deliberately. Such introductions reflect man's increasing use of the sea. In some cases, such as the use of *Crassostrea gigas* in European fisheries, introductions have been beneficial, but in others (for example various oyster pests and competitors, such as the slipper limpet *Crepidula fornicata*) this is not so. Introduction of alien species detrimental to resource use or potential should be regarded as a form of pollution. Any introduction should be regarded with extreme caution since, if it proves successful, it is bound to alter community structure, and native species, which might themselves have been a resource, may decline.

Further examples of introductions are the deliberate transfer of fish such as the striped bass *Morone saxatilus* from the eastern seaboard to the San Francisco Bay area of the United States, and the accidental arrival and subsequent hybridization of *Spartina* in Europe (Chapter 7). The latter has resulted in some displacement of *Zostera*, loss of feeding for birds and loss of recreational amenity for man. The Indo-Pacific diatom *Biddulphia sinensis* is common in the North Sea, and *Sargassum muticum*, 'Jap-weed', is established on the south coast of England where it is feared it will hamper small-boat navigation, sea angling and industrial cooling-water intake.

Apart from deliberate introduction the main invasion routes have been via shellfish importation, and bilgewater and fouling of ships. The initial establishment of the barnacle *Elminius* in north-western Europe appears to have occurred following temporary laying up of shipping from Australia in Europe at the onset of World War II. Both the Suez and Panama Canals link areas with different fauna and flora and could form important marine invasion routes. This has already occurred in the Mediterranean where several Red Sea fish and crustaceans are established.

There have been similar introductions to maritime land and fresh water. Examples are the pondweed *Elodea canadensis* which has been naturalized throughout much of Europe since the mid-nineteenth century, and, also in Europe, the establishment of the coypu, a rodent which can cause extensive damage to river-banks and dykes. Numerous European plants are established in North America, especially on the eastern seaboard, and large numbers of animals and plants from the Northern Hemisphere are now found in Australia and New Zealand (Elton, 1958).

Fouling

Any sea/solid interface acquires biota through settlement, growth and succession. Man-made structures are no exception; if the process is detrimental it is called fouling. The adverse effects may include corrosion, restriction of water movement and increased drag. Fouling usually begins with the adsorption of chemicals into the solid surface; the first biological stage is film formation by bacteria, often accompanied by diatoms and other protists. *Corrosion* is promoted by activity of sulphate-reducing bacteria in the inner film, or, for example, beneath the paintwork of ships' hulls. The H_2S produced induces formation of corrosion cells within the metal. Corrosion is often aided by surface irregularities which themselves can be brought about by impingement of fouling organisms against the surface. After the primary film stage other forms such as hydroids and thread-like algae (for example *Tubularia* and *Ectocarpus*) may settle. These are often succeeded by barnacles (such as *Elminius* and *Semibalanus*), sea-mats, serpulid worms, sea-squirts and mussels (*Mytilus*), as well as by larger algae.

Fouling within intake and pipe systems, such as power-station culverts and condensers and marine engine-cooling systems, can restrict water flow severely and cause considerable loss of efficiency and increase in corrosion. Fouling increases hull drag significantly; the British Admiralty has suggested a frictional resistance allowance of 0.25% per day at sea in temperate waters and 0.5% in the tropics. The latter would mean a 50% increase in fuel costs within three months. Fouling also increases water-movement stress on structures such as oil rigs. Hull fouling is countered by application of paints containing copper, tin or mercury compounds; pipe fouling is usually countered by chlorination. Wood- or rock-boring organisms also damage man-made structures. The best known examples are the so-called shipworms (actually bivalves) such as *Teredo* and *Bankia* and the gribble *Limnoria* (an isopod crustacean). In the coastal waters of British Columbia it is believed that *Bankia* causes logging-trade losses in excess of $1 m per year.

Access, recreation and tourism

This chapter is mainly concerned with man's use of tangible resources. Although the economic benefits accruing to local communities from tourism are readily quantifiable, the physical and spiritual pleasures gained by the tourists are not. Nevertheless for many people 'going on holiday' is synonymous with 'going to the seaside'. In Great Britain, coastal holiday towns were initially founded or grew up in the eighteenth century as health resorts (seawater spas) for the aristocracy and upper middle classes. The advent of coastal passenger steamers and railways in the mid-nineteenth century brought the resorts within reach of other social classes. 'The imprisoning of a previously rural population in industrial towns from which, until the establishment of the railways, escape was almost impossible, aroused strong subconscious desires for freedom .... On the shores of Britain was another world of time and space' (Anderson and Swinglehurst, 1978).

Similar forces were at work in much of the 'Western' world and, following the invention of the internal combustion engine and of aeroplanes, the seaside has become accessible to even more people at even greater distances. This has added to development pressures and to competition for coastal resources. The construction of better roads, harbours and airports to facilitate access and of marinas to attract boat owners are obvious examples. The economic value of tourism can be considerable; in Bermuda in 1978 about two-fifths of the gross domestic product and about two-thirds of foreign earnings were derived from tourism and more than a quarter of the working population were involved in the hotel and catering trade (see Hayward *et al.*, 1981). Coastal towns also often derive considerable economic benefit from people who retire to live in their vicinity. In some places the resulting social imbalance has caused concern. Visitor pressure may also cause ecological problems such as the erosion of clifftop vegetation, denudation of sand dunes, alterations of intertidal community structure (for example by trampling or by shellfish collection) and disturbance of bird flocks. Much further work on sociological and ecological interaction with recreation and tourism is required.

Planning and legislation

There has been a significant movement towards more integrated coastal planning in many countries since about 1970. For example, in the United States great impetus was given by the Coastal Zone Management Act of 1972 which required maritime states to produce an approved coastal management programme. In the United Kingdom, however, the general government response in attempting to regulate coastal activities has been to assign their management to existing departments or agencies. Some of the consequences have been

> (i) the absence of an integrated policy for the users of the coastal zone based upon an overall development plan for the entire zone; (ii) the absence of a single Ministry or public authority with overall or even primary responsibility for the coastal zone; (iii) the absence of uniform mechanisms whereby claims for new uses can be evaluated or priorities between different uses determined; (iv) serious uncertainty as to who has authority in any particular area, and especially the lack of a straightforward means of discovering who has ownership or lessee rights; (v) the regulatory legislation tends to deal with the control of individual uses and to be applicable to different strips of the zone; (vi) not infrequently legal regulation is the consequence of pressures from sectional interest groups and does not necessarily involve a satisfactory balance of competing uses (Anon, 1979).

In most of the British Isles the intertidal zone is state-(Crown)-owned, thus giving wide rights and traditions of free public access, but in many countries access has to be negotiated by government or other agencies. The differing national positions regarding shore rights are partly mirrored by the situation regarding the continental shelf and its waters. Under Roman law the seas (and seashores) were a freely accessible and communal asset and, consequently, not under state control. This is the concept of *res communis*. The opposing view, *res nullius*, was that since the seas were not originally owned by anyone they could be appropriated by sovereign powers. The latter doctrine became generally adopted for coastal waters because of their value for trade and fisheries (see Sharp and Stewart, 1978).

Many attempts to extend such territorial waters over a wider scale have been backed by armed power and have led to sea-battles, for example between Spain, Britain and the Netherlands. In 1972 there were 103 countries claiming territorial waters 12 miles or less offshore (several still only the 3 miles which was the distance seventeenth-century cannon could fire) and 14 claimed distances up to 200 miles. The latter is now being generally adopted as the 'exclusive economic zone'. The United Nations have consistently worked toward establishing international laws and conventions for use of the oceans. The greatest successes to date have been in relation to pollution from oil and from dumping and the adoption in 1982 of the Convention on the Law of the Sea by 119 countries (Borgese, 1983).

Coastal conservation

The preceding sections show that man has a major impact on the coastal zone but that he is concerned that this may be detrimental to his own interests. Conservation has been described succinctly as 'the wise use of resources'. Four main sources of conservation problems are (1) that some resource uses are incompatible, (2) that it is difficult to determine the priority of use, (3) ignorance and (4) previous wrong decisions. The most important of these is probably ignorance; without information, priorities cannot be determined and decisions will be based on arbitrary or personal criteria. The latter frequently means short-term gain at the expense of sustained long-term value.

Planning, conservation and statements of environmental impact are basically concerned with five main questions, namely (1) what are the resources, (2) what level of use will not adversely affect them, (3) what is the most desirable use or use-combination, (4) can this be achieved and (5) if so, how?

Detection and prediction of change

Biological communities exist in a state of dynamic flux. Natural change may be relatively predictable, as with seres and seasons, or it may be stochastic, as with storm damage. However, taken over sufficient time and area, a community exhibits stability unless long-term factors, such as climatic shift, are operating. Change can be detected only by sequential sampling. Occasionally this can be done retrospectively, for example through examination of the varying pollen content of peat layers. Since animals and plants incorporate material from their environment, it is sometimes possible to use them as sampling devices, for example for heavy metal detection; such work may give forewarning of pollution-induced community changes before they become catastrophic.

The most frequently used parameter to assess change in community 'health' is *species diversity*. At its simplest this is the number of different species within a community; however, there is obviously some difference between a given area containing for example fifteen species each with only one individual and the same area containing fifteen species represented by a hundred individuals. Thus species diversity indices taking account of *relative abundance* and/or the evenness of numerical distribution (*equitability*) have been derived. Two commonly-used indices are: (1) the *Shannon-Wiener index* in which diversity is expressed as e^H and

$$H = - \sum_{i=1}^{N} p_i \log_e p_i$$

where p_i is the proportion of species i in a sample of N species, and (2) *Gleason's index* where the diversity

$$d = \frac{(S-1)}{\log N}$$

where S is the number of species and N is the total number of organisms in a representative sample.

In general it is held that macrofaunal diversity decreases with increase in environmental stress (natural or anthropogenic) and thus that it can be used to indicate pollution. It has also been suggested that change in communities can be detected through plotting the cumulative number of species represented within geometric size classes of increasing abundance on a logarithmic scale, a normal distribution giving a straight line (the *log-normal* plot), the angle of which lessens if the community has been stressed (Gray and Mirza, 1979). This technique has been widely used but severely criticized (see Lambshead *et al.*, 1983) and it may be better to express the relationship of different species' abundances in a community in terms of *k-dominance* (Figure 11.6).

Another commonly used method of comparison is to construct or compute a faunal *similarity matrix* or *dendrogram*. In these, all samples in a series are compared with each other (Figure 11.7) and it can thus be determined whether there are any substantial community differences due for example to pollution or to sampling separate habitats. It has been

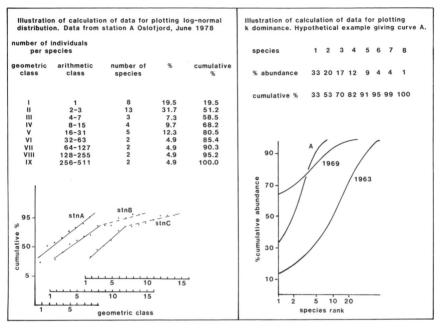

Figure 11.6 Comparison of benthic samples by log-normal distribution (from Gray and Mirza, 1979) and by k-dominance (after Lambshead et al., 1983). Oslofjord stations B and C are heavily polluted with organic matter. The k-dominance 1963 and 1969 curves are for Loch Eil before and after commencement of pulp-mill discharge in 1966.

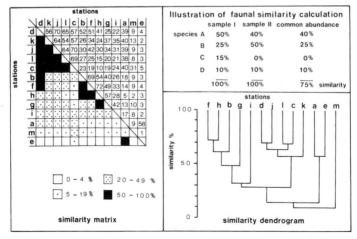

Figure 11.7 Comparison of benthic samples by similarity matrix and dendrogram. (After Pearson and Rosenberg, 1978.)

suggested that, rather than attempting to examine all species in a community, it would be better to concentrate on a few *key-species*, such as limpets or mussels, which are dominant in determining community structure, or on species or life-stages which are known to be pollution-sensitive. It is obviously best to detect pollution-induced changes before they become acute; this may be possible through study of enzyme-activity change within species (Blackstock, in Kinne and Bulnheim, 1980).

Coastal nature reserves

To understand the way in which man modifies coastal communities it is necessary to determine their composition and control mechanisms in as near natural conditions as possible. This and the protection of rare habitats and species are among the most important reasons for establishing nature reserves. Many reserves also have an important function in educating biologists and/or the general public as well as in research. The criteria for evaluating terrestrial sites for reserve declaration have been given as (1) extent—basically the bigger the better; (2) diversity—both in terms of species and habitats; (3) naturalness; (4) occurrence of rare species; (5) fragility—that is the ease with which the site might be damaged if unprotected; (6) representativeness—it is very important that a range of 'ordinary' natural areas are protected as well as those places of specialized interest; (7) recorded history—this is of value for assessment of change; (8) position—a criterion for ease of management; (9) potential—a site such as a disused quarry may be of value in that new habitat can be created there; and (10) intrinsic appeal—a site may be aesthetically pleasing.

The need for marine nature reserves has now been generally recognized. Similar reasoning can be applied to site selection but, since the majority of marine communities are considerably more pristine and more variable than those on land, the dominant criterion is usually that of extent (Anon.,1979). It has to be accepted that selective protection through reserve establishment will never apply to more than a very small percentage of the environment and that true conservation can only be achieved through much wider measures. Perhaps this is more obvious in the sea than on land, since marine communities are interlinked through factors such as life history (for example planktonic larvae of benthic animals), transfer of food (surface productivity fuels deeper communities) and the transfer of nutrients and pollutants by water movement. It is virtually impossible to protect a marine community by enclosure.

System modelling

Mathematics and ecological knowledge can be combined to aid understanding of community and ecosystem processes and to aid in the prediction of natural or man-induced change. Whereas individuals, populations, small communities and other ecosystem sub-units can be established or simulated with real organisms in laboratory conditions, this is well-nigh impossible for complete ecosystems. However, providing there is sufficient information, such systems can be modelled, for example by digital or analog computation (see Longhurst, 1981). Ecosystem models use the flow of energy or matter to link component parts (Figure 11.8). It is then often possible to simulate the interactions which would result from changes in the components and pathways. Some of the most extensive analog models of this kind have been developed for the Baltic (e.g. Jansson, 1972) and elsewhere using 'energy circuit language' (Figure 11.9).

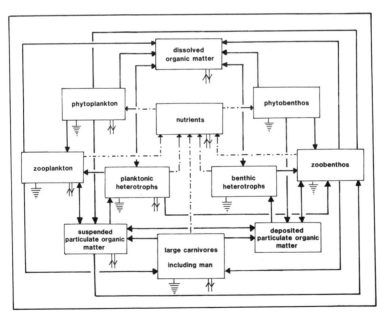

Figure 11.8 Flow diagram of carbon for a computer model (GEMBASE) of an estuarine ecosystem, earthing symbols represent loss of energy as heat. (After Longhurst and Radford, in Longhurst, 1981.)

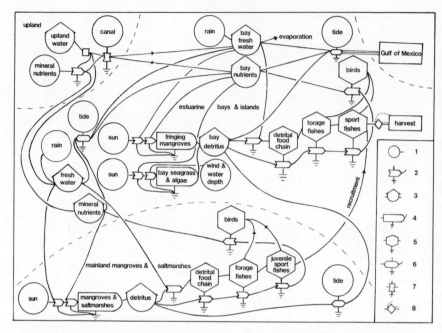

Figure 11.9 A model of energy flow in Odum's energy circuit language for a sub-tropical mangrove system and related coastal areas. Main symbols are: 1, energy source; 2, workgate (controlling energy flow enables another flow); 3, store of substance or energy; 4, plant population (energy is captured and partly recycled internally); 5, consumer unit (also partly recycling energy for maintenance); 6, two-way gate (similar to workgate but energy flow may be reversed); 7, switch (energy flow only possible if mechanism activated); 8, economic transaction (opposite cash and energy flows). (After Clark, in Kinne and Bulnheim, 1980)

In conclusion

Very few attempts have been made to study, let alone manage, entire coastal ecosystems (see Clark, in Kinne and Bulnheim, 1980). The difficulties are considerable especially when urban areas are included. Exploration by ecologists and economists of the multiple interactions between man and the coastal zone environment are still in a relatively early stage. It is hardly surprising therefore that many planners and legislators seem unaware of the need for an integrated approach. The authors hope that this book will help in the general process of coastal education, which itself must be part of the wider World Conservation Strategy whose three main objectives are (1) to maintain essential ecological processes and life support systems; (2) to preserve genetic diversity, which is being dangerously impoverished; and (3) to ensure the sustainable use by us and our children of species and ecosystems.

Envoi

'Knowledge must lead to action; intellectual understanding is useless unless linked to moral commitment' (Berry, in Anon, 1983).

REFERENCES AND FURTHER READING

Chapter 1

Allen, J.R.L. (1970) *Physical Processes of Sedimentation*. Allen and Unwin, London.

Barnes, R.S.K. (ed.) (1977) *The Coastline*. Wiley, Chichester.

Dring, M.J. (1982) *The Biology of Marine Plants*. Arnold, London.

Fairbridge, R.W. (ed.) (1966) *The Encyclopedia of Oceanography*. Van Nostrand Reinhold, New York.

Goldberg, E.D. (ed.) (1974) *The Sea: Ideas and Observations in Progress in the Study of the Seas*. vol. 5. Wiley-Interscience: New York.

Hill, M.N. (ed.) (1962-3) *The Sea: Ideas and Observations in Progress in the Study of the Seas*, vols. 1-3. Wiley-Interscience, New York.

Inman, D.L. (1949) Sorting of sediments in the light of fluid mechanics. *J. sediment. Petrol.* **19**, 51-70.

Mann, K.H. (1982) *Ecology of Coastal Waters: A Systems Approach*. Blackwell, Oxford.

Meadows, P.S. and Campbell, J.I. (1978) *An Introduction to Marine Science*. Blackie, Glasgow and London.

Parsons, T.R., Takahashi, M. and Hargrave, B. (1984) *Biological Oceanographic Processes* (3rd edn.). Pergamon, Oxford.

Postma, H. (1967) Marine pollution and sedimentology. In *Pollution and Marine Ecology*, Olson, T.A. and Burgess, T.J. (eds.), Wiley-Interscience, New York, 225-234.

Riedl, R. (1971) Water movement: animals. In *Marine Ecology*, Kinne, O (ed.), Vol. 1, Pt. 2, Wiley-Interscience, London, 1123-1149.

Riley, G.A. (1970) Particulate and organic matter in sea water. *Adv. Mar. Biol.* 8, 1-118.

Steers, J.A. (1971) *Applied Coastal Geomorphology*. Macmillan, London.

Steers, J.A. (1964) *The Coastline of England and Wales*. (2nd edn.). Cambridge University Press, Cambridge.

Stewart, M.G. (1979) Absorption of dissolved organic nutrients by marine invertebrates. *Oceanogr. Mar. Biol. Ann. Rev.* **17**, 163-192.

Strahler, A.N. (1960) *Physical Geography*. Wiley, New York.

Strahler, A.N. (1963) *The Earth Sciences*. Harper & Row, New York.

Turekian, K.K. (1976) *Oceans*. Prentice-Hall, New Jersey.

Chapter 2

Bowman, M.J. and Esias, W.E. (eds.) (1978) *Oceanic Fronts in Coastal Processes*. Springer, Berlin.

Conover, R.J. (1978) Feeding interactions in the pelagic zone. *Rapp. P.-v. Réun. Cons. int. Explor. Mar.* **173**, 66-76.

Cushing, D.H. (1975) *Marine Ecology and Fisheries*. Cambridge University Press, Cambridge.

Cushing, D.H. and Walsh, J.J. (eds.) (1976) *The Ecology of the Seas*. Blackwell, Oxford.

Dawes, C.J. (1981) *Marine Botany*. Wiley, New York.

Fogg, G.E. (1980) Phytoplanktonic primary production. In *Fundamentals of Aquatic Ecosystems*, Barnes, R.S.K. and Mann, K.H. (eds.). Blackwell, Oxford, 24-45.

Kremer, J.N. and Nixon, S.W. (1977) *A Coastal Marine Ecosystem Simulation and Analysis*. Springer, Berlin.

Landry, M.R. (1977) A review of important concepts in the trophic organization of pelagic ecosystems. *Helgol. wiss. Meeresunters.* **30**, 8–17.

Leadbetter, M. (1979) Langmuir circulations and plankton patchiness. *Ecol. Modelling* **7**, 289–310.

Livingston, R.J. (ed.) (1979) *Ecological Processes in Coastal and Marine Systems.* Plenum, New York.

Longhurst, A.R. (ed.) (1981) *Analysis of Marine Ecosystems.* Academic Press, London.

Margalef, R. (1962) Succession in marine populations. *Adv. Front. Pl. Sci.* **2**, 137–188.

Mann, K.H. (1982) *Ecology of Coastal Waters: A Systems Approach.* Blackwell, Oxford.

Parsons, T.R., Takahashi, M. and Hargrave, B. (1984) *Biological Oceanographic Processes.* Pergamon, New York.

Raymont, J.E.G. (1980) *Plankton and Productivity in the Oceans. 1. Phytoplankton.* Pergamon, Oxford.

Raymont, J.E.G. (1983) *Plankton and Productivity in the Oceans. 2. Zooplankton.* Pergamon, Oxford.

Riley, G.A. (1976) A model of plankton patchiness. *Limnol. Oceanogr.* **21**, 873–880.

Ryther, J.H. (1969) Photosynthesis and fish production in the sea. *Science* **166**, 72–76.

Ryther, J.H. and Dunstan, W.M. (1971) Nitrogen, phosphorus and eutrophication in the coastal marine environment. *Science* **171**, 1008–1012.

Sieburth, J.M. (1979) *Sea Microbes.* Oxford University Press, New York.

Smayda, T.A. (1970) The suspension and sinking of phytoplankton in the sea. *Oceanogr. Mar. Biol. Ann. Rev.* **8**, 353–414.

Steele, J.H. (ed.) (1978) *Spatial Pattern in Plankton Communities.* Plenum, New York.

Werner, D. (ed.) (1977) *The Biology of Diatoms.* Blackwell, Oxford.

Chapter 3

Connell, J.H. (1961) The influence of interspecific competition and other factors on the distribution of the barnacle *Chthalamus stellatus. Ecology* **42**, 710–723.

Connell, J.H. (1972) Community interactions on marine rocky intertidal shores. *Ann. Rev. Ecol. Syst.* **3**, 169–192.

Connell, J.H. and Keough, M.J. (1985) Disturbance and patch dynamics of subtidal marine animals on hard substrata. In *Natural Disturbance: an Evolutionary Perspective,* Picket, S.T.A. and White, P.S. (eds.) Academic Press, New York, in press.

Crisp, D.J. (1984) Overview of research on marine invertebrate larvae, 1940–1980. In *Marine Biodeterioration: An Interdisciplinary Study,* J.D. Costlow and Tipper, R.C. (eds.) Naval Institute Press, Annapolis, Maryland, 103–126.

Dayton, P.K. (1971) Competition, disturbance and community organization: the provision and subsequent utilization of space in a rocky intertidal community. *Ecol. Monogr.* **41**, 351–389.

Dayton, P.K. (1975) Experimental evaluation of ecological dominance in a rocky intertidal community. *Ecol. Monogr.* **45**, 137–159.

Dring, M.J. (1982) *The Biology of Marine Plants.* Arnold, London.

Ebling, F.J., Kitching, J.A., Muntz, L. and Taylor, C.M. (1964) The ecology of L. Ine 13. Experimental observations on the destruction of *Mytilus edulis* and *Nucella lapillus* by crabs. *J. Anim. Ecol.* **33**, 73–82.

Hawkins, S.J. and Hartnoll, R.G. (1983) Grazing of intertidal algae by marine invertebrates. *Oceanogr. Mar. Biol. Ann. Rev.* **21**, 195–282.

Harger, J.R. (1972) Competitive coexistence among intertidal invertebrates. *Amer. Sci.* **60**, 600–607.

Hiscock, K. and Mitchell, R. (1980) The description and classification of sublittoral epibenthic ecosystems. In *The Shore Environment, 2: Ecosystems,* J.H. Price, Irvine, D.E.G. and Farnham, W.F. (eds.) Academic Press, London, 323–370.

204 AN INTRODUCTION TO COASTAL ECOLOGY

Jackson, J.B.C. (1977) Competition on marine hard substrata: The adaptive significance of solitary and colonial strategies. *Am. Nat.* **111**, 743–767.

Jackson, J.B.C. (1983) Biological determinants of present and past sessile animal distributions. In *Biotic Interactions in Recent and Fossil Benthic Communities*, Tevesza, M.J.S. and McCall, P.L. (eds.) Plenum, New York, 39–119.

Lewis, J.R. (1964) *The Ecology of Rocky Shores*. English Universities Press, London.

Lewis, J.R. (1977). Rocky foreshores. *In The Coastline*, Barnes, R.S.K. (ed.), Wiley, London, 147–158.

Lubchenco, J. (1980) Algal zonation in the New England rocky intertidal community: An experimental analysis. *Ecology* **61**, 333–344.

Menge, B.A. and Lubchenco, J. (1981) Community organization in temperate and tropical rocky intertidal habitats: prey refuges in relation to consumer pressure gradients. *Ecol. Monogr.* **51**, 429–450.

Moore, P.G. and Seed, R. (eds.) (1985) *The Ecology of Rocky Coasts*. Hodder and Stoughton, Sevenoaks.

Paine, R.T. (1966) Food web complexity and species diversity. *Am. Nat.* **100**, 65–75.

Paine, R.T. (1977) Controlled manipulations in the marine intertidal zone and their contributions to ecological theory. *Acad. Natl. Sci. Phil. Spec. publ.* **12**, 245–270.

Paine, R.T. (1984) Ecological determinism in the competition for space. *Ecology* **65**, 1339–1348.

Paine, R.T., Castilla, J.C. and Cancino, J. (1985) Perturbation and recovery patterns of starfish dominated intertidal assemblages in Chile, New Zealand and Washington State. *Am. Nat.* (in press).

Peterson, C.H. (1979) The importance of predation and competition in organizing the intertidal epifaunal communities of Barnegat Inlet, New Jersey. *Oecologia* **39**, 1–24.

Schonbeck, M. and Norton, T.A. (1980) Factors controlling the lower limits of fucoid algae on the shore. *J. Exp. Mar. Biol. Ecol.* **43**, 131–150.

Seed, R. (1969) The ecology of *Mytilus edulis* (Lamellibranchiata) on exposed rocky shores. 2. Growth and mortality. *Oecologia* **3**, 317–350·

Seed, R. and O'Connor, R.J. (1981) Community organization in marine algal epifaunas. *Ann. Rev. Ecol. Syst.* **12**, 49–74.

Sousa, W.P. (1985) Disturbance and patch dynamics on rocky intertidal shores. In *Natural Disturbance: an Evolutionary Perspective*, Pickett, S.T.A. and White, P.S. (eds.). Academic Press, New York, in press.

Stephenson, T.A. and Stephenson, A. (1972) *Life Between Tidemarks on Rocky Shores*. Freeman, San Francisco.

Suchanek, T.H. (1981) The role of disturbance in the evolution of life history strategies in the intertidal mussels *Mytilus edulis* and *M. californianus*. *Oecologia* **50**, 143–152.

Underwood, A.J. (1978) The refutation of critical tidal levels as determinants of the structure of intertidal communities on British shores. *J. Exp. Mar. Biol. Ecol.* **33**, 261–276.

Underwood, A.J., Denley, E.J. and Moran, M.J. (1983) Experimental analyses of the structure and dynamics of mid shore rocky intertidal communities in New South Wales. *Oecologia* **56**, 202–219.

Vermeij, G.J. (1978) *Biogeography and Adaptation Patterns of Marine Life*. Harvard University Press, Cambridge, Mass.

Warner, G.F. (1984) *Diving and Marine Biology: The Ecology of the Sublittoral*. Cambridge University Press, Cambridge.

Wolcott, T.G. (1973) Physiological ecology and intertidal zonation in limpets (*Acmaea*): a critical look at 'limiting factors'. *Biol. Bull.* **145**, 389–422.

Chapter 4

Bell, S.S. and Coen, L.D. (1983) Investigations of meiofauna I. Abundances on repopulation of the tube caps of *Diopatra cuprea* (Polychaeta: Onuphidae) in a subtropical system. *Mar. Biol.* **67**, 303–309.

Boaden, P.J.S. (1977) Thiobiotic facts and fancies (Aspects of the distribution and evolution of anaerobic meiofauna). *Mikrofauna Meeresbod.* **61**, 45–63.

Briggs, J.C. (1974) *Marine Zoogeography.* McGraw-Hill, New York.

Earll, R. and Erwin, D.G. (eds.) (1981) *Sublittoral Ecology. The Ecology of the Shallow Sublittoral Benthos.* Clarendon Press, Oxford.

Ekman, S. (1953) *Zoogeography of the Sea.* Sidgwick and Jackson, London.

Fenchel, T. (1969) The ecology of marine microbenthos IV. Structure and function of the benthic ecosystem, its chemical and physical structure and the microfauna community with special reference to the ciliated Protozoa. *Ophelia* **6**, 1–182.

Fenchel, T. (1970) Studies in the decomposition of organic detritus derived from the turtle grass *Thalassia testudinum. Limnol. Oceanogr.* **15**, 14–20.

Fenchel, T. (1978) The ecology of micro- and meiobenthos. *Ann. Rev. Ecol. Syst.* **9**, 99–121.

Fenchel, T. and Riedl, R.J. (1970) The sulfide system: a new biotic community underneath the oxidized layer of marine sand bottoms. *Mar. Biol.* **7**, 255–268.

George, D. and George, J. (1979) *Marine Life. An Illustrated Encyclopedia of Invertebrates in the Sea.* Harrap, London.

Gerlach, S.-A. (1978) Food-chain relationships in subtidal silty and marine sediments and the role of meiofauna in stimulating bacterial productivity. *Oecologia* **33**, 55–69.

Gray, J.S. (1981) *The Ecology of Marine Sediments: An Introduction to the Structure and Function of Benthic Communities.* Cambridge University Press, Cambridge.

Hulberg, L.W. and Oliver, J.S. (1980) Caging manipulations in marine soft-bottom communities: importance of animal interactions on sedimentary habitat modification. *Can. J. Fish. Aquat. Sci.* **37**, 1130–1139.

Hylleberg, J. (1975) Selective feeding by *Abarenicola pacifica* with notes on *Abarenicola vagabunda* and a concept of gardening in lugworms. *Ophelia* **14**, 113–138.

Kohlmeyer, J. and Kohlmeyer, E. (1979) *Marine Mycology. The Higher Fungi.* Academic Press, New York.

Livingston, R.J. (ed.) (1979) *Ecological Processes in Coastal and Marine ecosystems.* Plenum, New York.

McLachlan, A. and Erasmus, T. (eds.) (1983) *Sandy Beaches as Ecosystems.* Junk, The Hague.

McRoy, C.P. and Helfferich, C. (eds.) (1977) *Seagrass Ecosystems. A Scientific Prespective.* Dekker, New York.

Mills, E.L. (1969) The community concept in marine zoology, with comments on continua and instability in some marine communities: a review. *J. fish. Res. Bd. Canada* **26**, 1415–1428.

Nedwell, D.B. and Brown, C.M. (eds.) (1982) *Sediment Microbiology.* Academic Press, London.

Price, J.H., Irvine, D.E.G. and Farnham, W.F. (eds.) (1980) *The Shore Environment, 2. Ecosystems.* Academic Press, London.

Reise, K. (1981) High abundance of small zoobenthos around biogenic structures in tidal sediments of the Wadden Sea. *Helgol wiss. Meeresunters.* **34**, 413–425.

Rhoads, D.C. (1974) Organism-sediment relations on the muddy sea floor. *Oceanogr. Mar. Biol. Ann. Rev.* **12**, 263–300.

Riedle, R.J. (1971) How much water passes through sandy beaches? *Int. Revue ges. Hydrobiol.* **56**, 923–946.

Riemann, F. and Schrager, M. (1978) The mucus trap hypothesis on feeding of aquatic nematodes and implications for biodegradation and sediment texture. *Oecologia.* **34**, 75–88.

Sieburth, J.M. (1979) *Sea Microbes.* Oxford University Press, New York.

Tenore, K.R. and Coull, B.C. (eds.) (1980) *Marine Benthic Dynamics.* University of South Carolina Press, Columbia.

Thorson, G. (1957) Bottom communities. In *Treatise on Marine Ecology and Paleoecology,* vol. 1: *Ecology* Hedgpeth, J.W. (ed.) *Mem. Geol. Soc. Amer.* **67** (1), 461–534.

Warwick, R.M., Joint, I.R. and Radford, P.S. (1979) Secondary production of the benthos in an estuarine environment. *Br. Ecol. Soc. Symp.* **19**, 429–450.

Wells, G.P. (1966) The lugworm (*Arenicola*)—a study in adaptation. *Neth. J. Sea Res.* 3, 294–313.
Wilson, K. (1983) Beach sediment temperature variations through depth and time. *Est. Coast. Shelf. Sci.* 17, 581–586.
Wood, E.J.F. (1965) *Marine Microbial Ecology.* Chapman and Hall, London.

Chapter 5

Barnes, R.S.K. (1984) *Estuarine Biology.* Edward Arnold, London.
Branch, M. and Branch, G. (1983) *The Living Shores of Southern Africa.* Struik, Cape Town.
Chapman, V.J. (ed.) (1977) *Ecosystems of the World.* 1. *Wet Coastal Ecosystems.* Elsevier, Amsterdam.
Duxbury, A.C. (1979) Upwelling and estuary flushing. *Limnol. Oceanogr.* 24, 627–633.
Dyer, K.R. (1973) *Estuaries: A Physical Introduction.* Wiley, London.
Green, J. (1968) *The Biology of Estuarine Animals.* Sidgwick and Jackson, London.
Haines, E.B. (1979) Interactions between Georgia salt marshes and coastal waters: a changing paradigm. In *Ecological Processes in Coastal and Marine Systems.*, R.J. Livingston (ed.) Plenum, New York, 35–46.
Hedgpeth, J.W. (1983) Brackish waters, estuaries and lagoons. In *Marine Ecology. A Comprehensive Integrated Treatise on Life in Oceans and Coastal Waters*, vol. 5, Kinne, O. (ed.) Wiley-Interscience, Chichester, 739–757.
Jones, N.V. and Wolff, W.J. (eds.) (1981) *Feeding and Survival Strategies of Estuarine Organisms.* Plenum, New York.
Kennedy, V.S. (ed.) (1982) *Estuarine Comparisons.* Academic, New York.
Ketchum, B.H. (ed.) (1983) *Estuaries and Enclosed Seas.* Elsevier, Amsterdam.
McLusky, D.S. (1981) *The Estuarine Ecosystem.* Blackie, Glasgow and London.
Perkins, E.J. (1974) *The Biology of Estuaries and Coastal Waters.* Academic Press, London.
Pritchard, D.W. (1967) Observations of circulation in coastal plain estuaries. In *Estuaries*, Lauff, G.H. (ed.), American Association for the Advancement of Science, Washington DC, 37–44.
Rankin, J.C. and Davenport, J.A. (1981) *Animal Osmoregulation.* Blackie, Glasgow and London.
Remane, A. and Schlieper, C. (1971) *Biology of Brackish Water.* Wiley-Interscience, New York.
Wangersky, P.J. (1977) The role of particulate matter in the productivity of surface waters. *Helgol. Wiss. Meeresunters* 30, 546–564.

Chapter 6

Abrams, P.A. (1984) Recruitment, lotteries and coexistence in coral reef fish. *Am. Nat.* 123, 44–55.
Bakus, G.J. (1981) Chemical defense mechanisms on the Great Barrier Reef, Australia. *Science* 211, 497–499.
Buss, L.W. and Jackson, J.B.C. (1979) Competitive networks: nontransitive competitive relationships in cryptic coral reef environments. *Am. Nat.* 113, 223–234.
Chappell, J. (1980) Coral morphology, diversity and reef growth. *Nature* 286, 249–252.
Connell, J.H. (1978) Diversity in tropical rain forests and coral reefs. *Science* 199, 1302–1310.
Endean, R. (1977) *Acanthaster planci* infestations of reefs of the Great Barrier Reef. *Proc. 3rd. Int. Coral Reef Symp.* 1, 185–191.
Glynn, P.W. (1976) Some physical and biological determinants of coral community structure in the Eastern Pacific. *Ecol. Monogr.* 46, 431–456.

Goreau, T.F., Goreau, N.I. and Goreau, T.J. (1979) Corals and coral reefs. *Sci. Amer.* **241**, 124–136.

Hughes, R.N. (1983) Evolutionary ecology of colonial reef-organisms, with particular reference to corals. *Biol. J. Linn. Soc.* **20**, 39–58.

Johannes, R.E., Wiebe, W.J., Crossland, C.J., Rimmer, D.W. and Smith, S.V. (1983) Latitudinal limits of coral reef growth. *Mar. Ecol. Prog. Ser.* **11**, 105–111.

Jones, O.A., and Endean, R. (eds.) *Biology and Ecology of Coral Reefs*, vol. 2 (1973), vol. 3 (1976), vol. 4 (1977), Academic Press, New York.

Lang. J.C. (1973) Interspecific aggression by scleractinian corals. 2. Why the race is not only to the swift. *Bull. Mar. Sci.* **23**, 260–279.

Lassig, B.R. (1977) Communication and coexistence in a coral reef community. *Mar. Biol.* **42**, 85–92.

Lewis, J.B. (1977) Processes of organic production on coral reefs. *Biol. Rev.* **52**, 305–347.

Lobel, P.S. (1980) Herbivory by damsel fishes and their role in coral reef community ecology. *Bull. Mar. Sci.* **30**, 273–289.

Muscatine, L. and Porter, J.W. (1977) Reef corals: Mutualistic symbioses adapted to nutrient poor environments. *Bioscience* **27**, 454–460.

Porter, J.W. (1976) Autotrophy, heterotrophy and resource partitioning in Caribbean reef-building corals. *Am. Nat.* **110**, 731–742.

Sale, P.F., McWilliam, P.S. and Anderson, D.T. (1978) Faunal relationships among the near-reef zooplankton at three locations on Heron Reef, Great Barrier Reef, and seasonal changes in this fauna. *Mar. Biol.* **49**, 133–145.

Sale, P.F. (1980) The ecology of fishes on coral reefs. *Oceanogr. Mar. Biol. Ann. Rev.* **18**, 364–421.

Sammarco, P.W. (1980) *Diadema* and its relationship to coral spat mortality: grazing, competition and biological disturbance. *J. Exp. Mar. Biol. Ecol.* **45**, 245–272'.

Sheppard, C.R.C. (1980) Coral cover, zonation and diversity on reef slopes of Chagos atolls, and population structures of the major species. *Mar. Ecol. Progr. Ser.* **2**, 193–205.

Stehli, F.G. and Wells, J.W. (1971) Diversity and age patterns in hermatypic corals. *Syst. Zool.* **20**, 115–126.

Stearn, C.W., Scoffin, T.P. and Martindale, W. (1977) Calcium carbonate budget of a fringing reef on the west coast of Barbados. *Bull. Mar. Sci.* **27**, 479–510.

Stoddart, D.R. and Yonge, C.M. (eds.) 1971. Regional Variation in Indian Ocean Coral Reefs (*Symp. zool. Soc. Lond.* **28**). Academic Press, London.

Yonge, C.M. (1963) The biology of coral reefs. *Adv. Mar. Biol.* **1**, 209–260.

Chapter 7

Ball, M.C. (1980) Patterns of secondary succession in a mangrove forest of Southern Florida. *Oecologia* **44**, 226–235.

Beeftink, W.G. (1977) Salt marshes. In *The Coastline*, Barnes, R.S.K. (ed.) Wiley, Chichester, 93–121.

Bell, S.S., Watzin, M.C. and Coull, B.C. (1978) Biogenic structure and its effect on the spatial heterogeneity of meiofauna in a salt marsh. *J. exp. mar. Biol. Ecol.* **35**, 99–107.

Bunt, J.S., Williams, W.T. and Clay, H.J. (1982) River water salinity and the distribution of mangrove species along several rivers in North Queensland. *Aust. J. Bot.* **30**, 401–412.

Chapman, V.J. (1960) *Salt Marshes and Salt Deserts of the World*. Hill, London.

Chapman, V.J. (1976) *Mangrove Vegetation*. Cramer, Vaduz.

Chapman, V.J. (ed.) (1977) *Wet Coastal Ecosystems*. Elsevier, Amsterdam.

Christensen, B. (1983) Mangroves—what are they worth? *Unasylva* **35**, 2–15.

Clark, J. (1977) *Coastal Ecosystem Management*. Wiley-Interscience, New York.

Clough, B.F. (ed.) (1982) *Mangrove Ecosystems in Australia: Structure, Function and Management*. Australian National University Press, Canberra.

Hamilton, P.Y. and Macdonald. K.B. (eds.) (1980) *Estuarine and Wetland Processes with Emphasis on Modelling.* Plenum, New York.

Howes, B.L., Howarth, R.W., Teal, J.M., and Valiela, I. (1981) Oxidation–reduction potentials in a salt marsh: spatial patterns and interactions with primary production. *Limnol. Oceanogr.* **26**, 350–360.

Jefferies, R.L. and Davies, A.J. (eds.) (1979) *Ecological Processes in Coastal Environments.* Blackwell, Oxford.

Jeffrey, D.W. (ed.) (1977) *North Bull Island, Dublin Bay—a Modern Coastal Natural History.* Royal Dublin Society, Dublin.

Lassere, P. and Postma, H. (eds.) (1982) Coastal Lagoons. *Oceanol. Acta.* **5**, suppl. 4.

Lawton, J.R., Todd, A. and Naidoo, D.K. (1981) Preliminary investigations into the structure of the roots of the mangroves, *Avicennia marina* and *Bruguiera gymnorrhiza,* in relation to ion uptake. *New Phytol.* **81**, 713–722.

Letzsch, W.S. and Frey, R.W. (1980) Erosion of salt marsh tidal creek banks. Sapelo Island, Georgia. *Senckenb. Marit.* **12**, 201–212.

Livingston, R.J. (ed.) (1979) *Ecological Processes in Coastal and Marine Systems.* Plenum, London.

Long, S.P. and Mason, C.F. (1983) *Saltmarsh Ecology.* Blackie, Glasgow, and London.

Lugo, A.E. and Snedaker, S.C. (1974) The ecology of mangroves. *Ann. Rev. Ecol. Syst.* **5**, 39–64.

McCoy, E.D. and Heck, K.L. (1976) Biogeography of corals, seagrasses and mangroves: an alternative to the centre of origin concept. *Syst. Zool.* **25**, 201–210.

McNae, W. (1968) A general account of the fauna and flora of mangrove swamps and forests in the Indo-West Pacific region. *Adv. Mar. Biol.* **6**, 73–270.

Nakanishi, H. (1981) Distribution and ecology of a *Paliurus ramosissimus* community. *Acta Phytotaxon. Geobot.* **32**, 105–113. (In Japanese).

Odum, W.E. and Heald, E.J. (1975) The detritus-based food web of an estuarine mangrove community. In *Estuarine Research,* vol. 1, Cronin, L.E. (ed.) Academic Press, New York, 265–286.

Perkins, E.J. (1974) *The Biology of Estuaries and Coastal Waters.* Academic Press, London.

Polderman, P.J.G. (1976) Seasonal aspects of algal communities in salt marshes. In *La végétation des vases salées.* Gehu, J.M. (ed.) *Ass. int. Phytosociologie Colloq. Phytosociologiques* **4**, Strauss-Cramer: Hirschberg, 479–487.

Rabinowitz, D. (1978) Mortality and initial propagule size in mangrove seedlings in Panama. *J. Ecol.* **66**, 45–52.

Ranwell, D.S. (1967) World resources of *Spartina townsendii* (*sensu lato*) and economic use of *Spartina* marshland. *J. Appl. Ecol.* **4**, 239–256.

Ranwell, D.S. (1972) *Ecology of Salt Marshes and Sand Dunes.* Chapman and Hall, London.

Reinold, R.J. and Queen, W.H. (eds.) (1974) *Ecology of Halophytes.* Academic Press, New York.

Rublee, P.A., Merkel, S.M. and Faust, M.A. (1983) The transport of bacteria in the sediments of a temperate marsh. *Est. Coast. Shelf Sci.* **16**, 501–509.

Scholander, P.F. (1968) How mangroves desalinate water. *Physiol. Plant.* **21**, 251–261.

Semenuik, V. (1983) Mangrove distribution in North-western Australia in relation to regional and local freshwater seepage. *Vegetatio* **53**, 11–31.

Smith, T.J. III (1982) Alteration of salt marsh plant community composition by grazing snow geese. *Holarct. Ecol.* **6**, 204–210.

Teal, J.M. (1962) Energy flow in the salt marsh ecosystem of Georgia. *Ecology* **43**, 614–624.

Thom, B.G. (1967) Mangrove ecology and deltaic geomorphology: Tabasco, Mexico. *J. Ecol.* **55**, 301–343.

Waisel, Y. (1972) *Biology of Halophytes.* Academic Press, New York.

Zedler, J.B. (1982) Salt marsh algal mat composition: spatial and temporal comparisons. *Bull. South. Calif. Acad. Sci.* **81**, 41–50.

Chapter 8

Barnes, R.S.K. (ed.) (1977) *The Coastline*. Wiley, Chichester.

Bramwell, M. (ed.) (1980) *The Atlas of World Wildlife*. Mitchell Beazley, London.

Clarke, A. (1983) Life in cold water: the physiological ecology of polar marine ectotherms. *Oceanogr. Mar. Biol. Ann. Rev.* **21**, 341–457.

Holdgate, M.W. (ed.) (1970) *Antarctic Ecology*. Academic Press, London.

Jehne, W. and Thompson, C.H. (1981) Endomycorrhizae in plant colonization on coastal sand-dunes at Cooloola, Queensland. *Aust. J. Ecol.* **6**, 221–230.

Laws, R.M. (ed.) (1984) *Antarctic Ecology* (vols. 1 and II). Academic Press, London.

Oosting, H.J. (1954) Ecological processes and vegetation of the maritime strand in the United States. *Bot. Rev.* **20**, 226–262.

Polunin, N.V.C. (1983) The marine resources of Indonesia. *Oceanogr. Mar. Biol. Ann. Rev.* **21**, 455–531.

Ranwell, D.S. (1972) *Ecology of Salt Marshes and Sand Dunes*. Chapman and Hall, London.

Remmert, H. (1964) Distribution and the ecological factors controlling distribution of the European wrackfauna. *Botanica Gothoburg.* **3**, 179–184.

Taylor, R.J. (1981) Shoreline vegetation of the Arctic Alaska Coast. *Arctic* **34**, 37–42.

Timms, B.V. (1982) Coastal dune waterbodies of north-eastern New South Wales. *Aust. J. mar. freshw. Res.* **32**, 203–222.

Chapter 9

Anon. (1983) *The Environmental Impact of Aquaculture*. Swedish Council for Planning and Coordination of Research, Stockholm.

Arntz, W.E. (1978) The 'upper part' of the benthic food web: the role of macrobenthos in the Western Baltic. *Rapp. P.-v. Réun. Cons. int. Explor. Mer.* **173**, 85–100.

Bone, Q. and Marshall, N.B. (1982) *Biology of Fishes*. Blackie, Glasgow and London.

Chapman, V.J. and Chapman, D.J. (1980) *Seaweeds and their Uses* (3rd edn.). Chapman and Hall, London.

Cheng, T.-H. (1968) Production of kelp: a major aspect of China's exploitation of the sea. *Econ. Bot.* **23**, 215–236.

Cushing, D.H. (1968) *Fisheries Biology. A Study in Population Dynamics*. University of Wisconsin, Madison.

Dayton, P.K. (1975) Experimental studies of algal canopy interactions in a sea otter-dominated kelp community at Amchitka Island, Alaska. *Fish. Bull.* **73**, 230–237.

Department of Agriculture (1984) *Report on the Sea and Inland Fisheries of Northern Ireland 1983*. Her Majesty's Stationery Office, Belfast.

Dring, M.J. (1982) *The Biology of Marine Plants*. Edward Arnold, London.

Edwards, E. (1979) *The Edible Crab and its Fishery in British Waters*. Fishing News Books, Farnham.

Estes, J.A. and Palmisano, J.F. (1974) Sea otters: their role in structuring nearshore communities. *Science* **285**, 1058–1060.

Firth, F.E. (ed.) (1969) *The Encyclopedia of Marine Resources*. Van Nostrand Reinhold, New York.

Food and Agricultural Organization (1984) *Yearbook of Fisheries Statistics. Catches and Landings for 1982*. vol. 54. FAO, Rome.

Gibson, R.N. (1982) Recent studies on the biology of intertidal fish. *Oceanogr. Mar. Biol. Ann. Rev.* **20**, 363–414.

Goldman, B. and Talbot, E.H. (1976) Aspects of ecology of coral reef fishes. In *Biology and Geology of Coral Reefs*, Jones, O.A. and Endean, R. (eds.), vol. 3, Academic Press, New York, 125–154.

Harden Jones, F.R. (1974) *Sea Fisheries Research*. Elek, London.
Jensen, A. and Stein, J.R. (eds.) (1979) *Proceedings of the Ninth International Seaweed Symposium*. Science Press, Princeton.
Kautsky, N. and Wallentius, I. (1980) Nutrient release from a Baltic *Mytilus* red algal community and its role in benthic and pelagic productivity. *Ophelia* **1** (suppl.), 17–30.
Kikuchi, T. (1974) Japanese contributions on consumer ecology in eelgrass (*Zostera marina* L.) beds, with special reference to trophic relations and resources in inshore fisheries. *Aquaculture* **4**, 145–160.
Kinne, O. (ed.) (1983) *Marine Ecology*. Vol. 5, pt. 2, *Ecosystems and Organic Resources*. Wiley-Interscience, Chichester.
Kinne, O., and Bulnheim, H.–P. (eds.) (1980) Protection of Life in the Sea. 14th European Marine Biology Symposium. *Helgol. Wiss. Meeresunters.* **33**, 1–772.
McHugh, J.L. (1976) Estuarine fisheries: are they doomed? In *Estuarine Processes*, vol. 1, Wiley, M. (ed.) Academic Press, New York, 15–27.
May, R.M., Beddington, J.R., Clark, C.W., Holt, S.J. and Laws, R.M. (1979) The management of multispecies fisheries. *Science* **205**, 267–277.
Ogden, J.C. and Lobel, P.S. (1978) The role of herbivorous fishes and urchins in coral reef communities. *Env. Biol. Fish* **3**, 49–63.
Ridgway, S.H. and Harrison, R.J. (1981) *Handbook of Marine Mammals*, vols. 1 and 2. Academic Press, London.
Tiews, K. (1969) Tagging of 60,000 common shrimps *Crangon crangon* (L.) and its results *Arch. Fischereiswiss* **20**, 33–41. (In German).
Wanders, J.B.W. (1977) The role of benthic algae in the shallow reef of Curacao (Netherlands Antilles) III: The significance of grazing. *Aquat. Bot.* **3**, 357–390.
Wharton, W.G. and Mann, K.H. (1981) The relationships between destructive grazing by sea-urchin *Strongylocentrotus droebachiensis* and the abundance of the American lobster *Homarus americanus* on the Atlantic Coast of Nova Scotia. *Can. J. Fish aquat. Sci.* **38**, 1339–1349.
Wheaton, F.W. (1977) *Aquacultural Engineering*. Wiley-Interscience, New York.
Wiborg, K.F. (1946) Research on the horse-mussel (*Modiola modiolus* (L.)) I. General biology, growth and economic significance. *Rep. Norw. Fishery. mar. Invest.* **8**, 1–85. (In Norwegian).

Chapter 10

Ashmole, N.P. (1971) Seabird ecology and the marine environment. In *Avian Biology*, Farner, D.S. and King, J.R. (eds.), vol. 1, Academic Press, New York, 223–286.
Branch, M. and Branch, G. (1983) *The Living Shores of Southern Africa*. Struik, Cape Town.
Brown, R.G.B. (1981) Seabirds at sea. *Oceanus* **24**, 31–38.
Burger, J. and Olla, B.L. (eds.) (1984) *Behaviour of Marine Animals. 5: Shorebirds: Breeding Behaviour and Populations*. Plenum, New York.
Burger, J., and Olla, B.L. (eds.) (1984) *Behaviour of Marine Animals. 6: Shorebirds: Migration and Foraging Behaviour*. Plenum, New York.
Burger, J., Olla, B.I. and Winn, H.E. (eds.) (1980) *Behaviour of Marine Animals. 4: Marine Birds*. Plenum, New York.
Dorst, J. (1974) *The Life of Birds* (vol. 21). Weidenfeld and Nicolson, London.
Evans, P.R. (1979) Adaptations shown by foraging shorebirds to cyclical variations in the activity and availability of their intertidal invertebrate prey. In *Cyclical Phenomena in Marine Plants and Animals.*, Naylor, E. and Hartnoll, R.G. (eds.) Pergamon, Oxford, 357–366.
Evans, P.R., Goss-Custard, J.D. and Hale, W.G. (eds.) (1984) *Coastal Waders and Wildfowl in Winter*. Cambridge University Press, Cambridge.
Feare, C.J. and Summers, R. (1985) Birds as predators on rocky shores. In *The Ecology of Rocky Coasts*, Moore, P.G., and Seed, R. (eds.) Hodder and Stoughton, Sevenoaks,

Furness, R.W. (1982) Competition between fisheries and seabird communities. *Adv. Mar. Biol.* **20**, 225–307.

Goss-Custard, J.D. (1977) Predator responses and prey mortality in the redshank *Tringa totanus* (L.) and a preferred prey *Corophium volutator* (Pallas). *J. Anim. Ecol.* **46**, 21–36.

Goss-Custard, J.D. (1980) Competition for food and interference among waders. *Ardea* **68**, 31–52.

Hale, W.G. (1980) *Waders.* Collins, London.

Lack, D. (1968) *Ecological Adaptations for Breeding in Birds.* Methuen, London.

Milne, H. and Dunnet, G.M. (1972) Standing crop, productivity and trophic relations of the fauna of the Ythan Estuary. In *The Estuarine Environment*, Barnes, R.S.K. and Green, J. (eds.) Applied Science, London, 86–106.

Nelson, B. (1980) *Seabirds: their Biology and Ecology.* Hamlyn, London.

O'Connor, R.J. (1981) Patterns of shorebird feeding. In *Estuary Birds of Britain and Ireland*, Prater, A.J. (ed.), Poyser, Calton, 34–50.

O'Connor, R.J. and Brown, R.A. (1977) Prey depletion and foraging strategy in the oystercatcher (*Haematopus ostralegus*). *Oecologia* **27**, 75–92.

Prater, A.J. (1972) The ecology of Morecambe Bay. 3. The food and feeding habits of the knot (*Calidris canutus* L.) in Morecambe Bay. *J. Appl. Ecol.* **9**, 179–194.

Rosa, N. (1978) What is a seabird? *Oceanus* **11**, 22–26.

Schneider, D.C. (1978) Equalisation of prey numbers by migratory shorebirds. *Nature* **271**, 353–354.

Storer, R.W. (1971) Adaptive radiation in birds. In *Avian Biology*, Farner, D.S. and King, J.R. (eds.), vol. 1, Academic Press, New York, 149–188.

Whittow, G.C. and Rahn, H. (eds.) (1984) *Seabird Energetics.* Plenum, New York.

Chapter 11

Anderson, J. and Swinglehurst, E. (1978) *The Victorian and Edwardian Seaside.* Country Life Books, London.

Anon. (1978) *Coastal Zone 1978 Symposium on Technical, Environmental, Socioeconomic and Regulatory Aspects of Coastal Zone Management* (Vols. I-III). American Society of Civil Engineers, New York.

Anon. (1979) *Nature Conservation in the Marine Environment.* Nature Conservancy Council, London, and Natural Environment Research Council, London.

Anon. (1981) *Atlas of the Seas around the British Isles.* Ministry of Agriculture, Fisheries and Food, Lowestoft.

Anon. (1983) *The Conservation and Development Programme for the U.K. A response to the World Conservation Strategy.* Kogan Page, London.

Borgese, E.M. (1983) The Law of the Sea. *Sci. Amer.* **248**, 28–35.

Brahtz, J.F.P. (ed.) (1972) *Coastal Zone Management: Multiple Use with Conservation.* Wiley, New York.

Briggs, J.C. (1974) *Marine Zoogeography.* McGraw-Hill, London.

Cotillon, J. (1978) La Rance Tidal Power Station, Review and Comments. In *Tidal Power and Estuary Management, Proc. 13th Symp.* Colston Res. Soc. Severn, R.I., Dinely, D.L. and Hawker, L.E. (eds.) Scientechnica, Bristol, 49–66.

Countryside Commission (1970) *The Planning of the Coastline.* Her Majesty's Stationery Office, London.

Elton, C.S. (1958) *The Ecology of Invasions by Animals and Plants.* Methuen, London.

Gerlach, S.A. (1981) *Marine Pollution.* Springer, Berlin.

Gray, J.S. and Mirza, F.B. (1979) A possible method for detecting pollution-induced disturbances in marine benthic communities. *Mar. Pollut. Bull.* **10**, 142–146.

Hayward, S.J., Gomez, V.H. and Sterrer, W. (eds.) (1981) *Bermuda's Delicate Balance. People and Environment.* Bermuda National Trust, Hamilton.

Jansson, B.-O. (1972) Ecosystem Approach to the Baltic Problem. *Bull. Ecol. Res. Comm.* **16**, Swedish Natural Science Research Council: Stockholm.

Kinne, O. (ed.) (1982-4) *Marine Ecology. A Comprehensive, Integrated Treatise on Life in Oceans and Coastal Waters* (vol. 5). Wiley-Interscience, Chichester.

Kinne, O. and Bulnheim, H.-P. (eds.) (1980) Protection of Life in the Sea. 14th European Marine Biology Symposium. *Helgol. Wiss Meeresunters.* **33**, 1–772.

Kullenberg, G. (ed.) (1982) *Pollutant Transfer and Transport in the Sea* (vols. 1 and II). CRC Press, Boca Raton.

Lambshead, P.J.D., Platt, H.M. and Shaw, K.M. (1983) The detection of differences among assemblages of marine benthic species based on an assessment of dominance and diversity. *J. nat. Hist.* **17**, 859–874.

Longhurst, A.R. (ed.) (1981) *Analysis of Marine Ecosystems.* Academic Press, London.

Mensah, T.A. (1984) Environmental protection: International approaches. *Mar. Policy* **8**: 95–105.

Mitchell, R. and Probert, P.K. (1981) *Severn Tidal Power—The Natural Environment.* Nature Conservancy Council, London.

Moreno, C.A., Sutherland, J.P. and Jara, H.F. (1984) Man as a predator in the intertidal zone of southern Chile. *Oikos* **42**, 155–160.

Nelson-Smith, A. (1972) *Oil Pollution and Marine Ecology.* Elek, London.

Pearson, T.H. and Rosenberg, R. (1978) Macrobenthic succession in relation to organic enrichment and pollution of the marine environment. *Oceanogr. Mar. Biol. Ann. Rev.* **16**, 229–311.

Sharp, D. and Stewart, J. (eds.) (1978) *Oceanography. Law of the Sea.* Open University, Milton Keynes.

Shaw, T.L. (ed.) (1980) *An Environmental Appraisal of Tidal Power Stations with Particular Reference to the Severn Barrage.* Pitman, London.

Van Der Heyden, A.A.M. and Scullard, H.H. (eds.) (1959) *Atlas of the Classical World.* Nelson, London.

Zijlstra, J.J. (1972) On the importance of the Waddenzee as a nursery area in relation to the conservation of the southern North Sea fishery resources. *Symp. Zool. Soc. Lond.* **29**, 233–258.

Index